Student Study Guide
Volume III
for
Tipler and Mosca's
Physics for Scientists and Engineers
Sixth Edition

Todd Ruskell
Colorado School of Mines

W.H. Freeman and Company
New York

Acknowledgments

I must first thank Paul Tipler and Gene Mosca for putting together an excellent textbook in the sixth edition of *Physics for Scientists and Engineers*. It has been a delight to work with. I must also thank Gene for his work on earlier versions of this study guide, which have been drawn on heavily.

I am indebted to the reviewers of this study guide: Elizabeth Behrman (Wichita State), Daniel Dale (University of Wyoming), Linnea Hess (Olympic College), and Oren Quist (South Dakota State). Their careful reading and insightful comments made this a much better study guide than it would have been otherwise. In addition, their thorough reviews helped uncover many errors that would have remained to be discovered by the users of this study guide. Their assistance is greatly appreciated. In spite of our combined best efforts, there may still be an occasional error in this study guide, and for those I assume full responsibility. Should you find errors or would like to bring another matter regarding this study guide to my attention, please do not hesitate to send them to me by using asktipler@whfreeman.com.

My wife Susan and daughter Allison provided immense support with their patience and understanding throughout the entire process of writing this study guide.

It was a pleasure to work with Susan Brennan, Clancy Marshall, Kharissia Pettus, and Kathryn Treadway who guided me through the creation of this study guide.

October, 2007

Todd Ruskell
Colorado School of Mines

Printed in the United States of America

ISBN: 1-4292-0411-7 (Volume 3: Chapters 34–41)

First Printing 2008

W. H. Freeman and Company
41 Madison Avenue
New York, NY 10010
Houndmills, Basingstoke
RG21 6XS, England
www.whfreeman.com

Contents

To the Student

This study guide was written to help you master Chapters 34 through 41 of Paul Tipler and Eugene Mosca's *Physics for Scientists and Engineers*, Sixth Edition. Each chapter of the study guide is divided into sections that match the textbook, and culminates in a short quiz designed to test your mastery of the subject. Each of these sections may contain the subsections below.

In a Nutshell: A brief overview of the important concepts presented in the section. This section is designed only to remind you of the key ideas. If a concept is not clear, you should refer back to the text for more detailed explanations and derivations.

Physical Quantities and Their Units: A list of the constants, units, and physical quantities introduced in the section.

Fundamental Equations: A list of fundamental equations introduced in the section. These expressions provide underpinning for the Important Derived Results.

Important Derived Results: In many sections the Fundamental Equations are applied to specific physical situations. This application can result in important derived results that apply to only those specific situations. These results are listed here.

Common Pitfalls: Warnings about commonly-made mistakes. In addition, there are conceptual questions designed to test your understanding of the physical principles discussed in the section.

Try It Yourself: Workbook style questions following the structure of the solutions to the worked Examples in the text. Final answers, with units, are provided, but it is up to you to use the space provided to fill in the required work for the intermediate steps. Most of these questions also have a Taking It Further question designed to enhance your understanding of and ability to interpret the problem's solution. You should answer these questions in the space provided before looking at the answers in the back of the study guide.

What Is the Best Way to Study Physics?

Of course there isn't a single answer to that. It is clear, however, that you should begin early in the course to develop the methods that work best for you. The important thing is to find the system that is most comfortable and effective for you, and then stick to it.

In this course you will be introduced to numerous concepts. It is important that you take the time to be sure you understand each of them. You will have mastered a concept when you fully understand its relationships with other concepts. Some concepts will seem to contradict other concepts or even your observations of the physical world. Many of the questions in this study guide are intended to test your understanding of concepts. If you find that your understanding of an idea is incomplete, don't give up; pursue it until it becomes clear. We recommend that you keep a list of the things that you come across in your studies that you do not understand. Then, when you come to understand an idea, remove it from your list. After you complete your study of each chapter, bring your list to your most important resource, your physics instructor, and ask for assistance. If you go to your instructor with a few well-defined questions, you will very likely be able to remove any remaining items from your list.

Like the Example problems presented in the textbook, the problem solutions presented in this study guide start with basic concepts, not with formulas. We encourage you to follow this practice. Physics is a collection of interrelated basic concepts, not a seemingly infinite list of disconnected, highly specific formulas. Although at times it may seem we present long lists of formulas, do not try to memorize long lists of specific formulas and then use these formulas as the starting point for solving problems. Instead, focus on the concepts first and be sure that you understand the ideas before you apply the formulas.

Probably the most rewarding (but challenging) aspect of studying physics is learning how to apply the fundamental concepts to specific problems. At some point you are likely to think, "I understand the theory, but I just can't do the problems." If you can't do the problems, however, you probably don't understand the theory. Until the physical concepts and the mathematical equations become your tools to apply at will to specific physical situations, you haven't really learned them. There are two major aspects involved in learning to solve problems: drill and skill. By drill we mean going through a lot of problems that involve the direct application of a particular concept until you start to feel familiar with the way it applies to different physical situations. Each chapter of the text contains about 35 single-concept problems for you to use as drill. Do a lot of these—at least as many as you need in order to feel comfortable handling them.

By skill we mean the ability both to recognize which concepts are involved in more advanced, multi-concept problems, and to apply those concepts to particular situations. The text has several intermediate-level and advanced-level problems that go beyond the direct application of a single concept. As you develop this skill you will master the material and become empowered. As you find that you can deal with more complex problems—even some of the advanced-level ones—you will gain confidence and enjoy applying your new skills. The examples in the textbook and the problems in this study guide are designed to provide you with a pathway from the single-concept to the intermediate-level and advanced-level problems.

A typical physics problem describes a physical situation—such as a child swinging on a swing—and asks related questions. For example: If the speed of the child is 5.0 m/s at the bottom of her arc, what is the maximum height the child will reach? Solving such problems requires you to apply the concepts of physics to the physical situation, to generate mathematical relations, and to solve for the desired quantities. The problems presented here and in your textbook are exemplars; that is, they are examples that deserve imitation. When you master the methodology presented in the worked-out examples, you should be able to solve problems about a wide variety of physical situations.

A good way to test your understanding of a specific solution is to take a sheet of paper, and—without looking at the worked-out solution of an Example problem—reproduce it. If you get stuck and need to refer to the presented solution, do so. But then take a fresh sheet of paper, start from the beginning, and reproduce the entire solution. This may seem tedious at first, but it does pay off.

This is not to suggest that you reproduce solutions by rote memorization, but that you reproduce them by drawing on your understanding of the relationships involved. By reproducing a solution in its entirety, you will verify for yourself that you have mastered a particular example problem. As you repeat this process with other examples, you will build your very own personal base of physics knowledge, a base of knowledge relating occurrences in the world around you—the physical universe—and the concepts of physics. The more complete the knowledge base that you build, the more success you will have in physics.

You should budget time to study physics on a regular, preferably daily, basis. Plan your study schedule with your course schedule in mind. One benefit of this approach is that when you study on a regular basis, more information is likely to be transferred to your long-term memory than when you are obliged to cram. Another benefit of studying on a regular basis is that you will get much more from lectures. Because you will have already studied some of the material presented, the lectures will seem more relevant to you. In fact, you should try to familiarize yourself with each chapter before it is covered in class. An effective way to do this is first to read the In a Nutshell subsections of that study guide chapter. Then thumb through the textbook chapter, reading the headings and examining the illustrations. By orienting yourself to a topic before it is covered in class, you will have created a receptive environment for encoding and storing in your memory the material you will be learning.

Another way to enhance your learning is to explain something to a fellow student. It is well known that the best way to learn something is to teach it. That is because in attempting to articulate a concept or procedure, you must first arrange the relevant ideas in a logical sequence. In addition, a dialog with another person may help you to consider things from a different perspective. After you have studied a section of a chapter, discuss the material with another student and see if you can explain what you have learned.

Chapter 34

Wave–Particle Duality and Quantum Physics

34.1 Waves and Particles

In a Nutshell

The first part of the twentieth century was an exceptionally creative time in physics. Einstein developed the theory of special relativity in 1905 and the theory of general relativity in 1916. The ideas of quantum physics began in 1900 with Planck's explanation of blackbody radiation. Einstein extended Planck's quantum ideas to explain the photoelectric effect in 1905, the same year he put forth the theory of special relativity. Ideas built upon ideas about how nature behaves. The fields of quantum physics, relativity, cosmology, and other areas of modern physics were given birth and nurtured by Einstein, Planck, Bohr, Fermi, Dirac, and many others, most of whom were awarded Nobel prizes for their work.

In our studies so far, we have seen that particles, in the absence of external forces, travel in straight lines, even when they travel through apertures. Particles exchange energy in collisions that occur at specific points in space and time.

Waves on the other hand, diffract, bending around corners and spreading out when they encounter small apertures. Two or more waves can interfere with each other, resulting in interesting distributions of energy. The energy of waves is spread out in space and deposited continuously as the wavefronts interact with matter.

If a wave travels only through large apertures relative to its wavelength λ, however, its propagation is indistinguishable from that of a beam of particles. Diffraction effects are negligible and interference fringes occur so close together we cannot distinguish them. In other words, light seems to travel only in straight lines. The result is that light appears to hit a detector like a stream of continuous particles which hit in such rapid succession we cannot measure the impact of each individual particle, only the collective result.

In the rest of this chapter we will examine the merging of the properties of particles and waves.

34.2 Light: From Newton to Maxwell

In a Nutshell

Isaac Newton (1642–1727) believed that light was composed of particles, a belief that was accepted for over a century. However, many experiments conducted in the 1800s suggested that Newton's particle picture of light was incorrect. Thomas Young in 1801 passed light from a single source through two closely spaced parallel slits and produced an interference pattern on a screen. Augustin Fresnel (1788–1827) developed the mathematical theory of light based on a wave picture, and showed that all experimental results agreed with his wave analysis.

James Clerk Maxwell in 1860 put forth his theory of electromagnetism, which predicted the existence of electromagnetic waves that propagate at the speed of light. The implication is that light is an electromagnetic wave, differing from other electromagnetic waves like television and radio only in wavelength and frequency. The wavelength of electromagnetic light waves is in the visible region of about 400 nm to about 700 nm (1 nm = 10^{-9} m). These and many other experiments seemed to show conclusively that light is a wave and not a particle.

34.3 The Particle Nature of Light: Photons

In a Nutshell

At the beginning of the 1900s, various experiments were performed that could not be explained with a wave picture of light. Instead, the experiments suggested that light energy comes in discrete amounts carried by localized packets called **photons.** Einstein, in explaining the photoelectric effect, first put forth the idea that each photon has an energy $E = hf = hc/\lambda$, where f is the frequency of light associated with each photon and $h = 6.626 \times 10^{-34}$ J·s $= 4.136 \times 10^{-15}$ eV·s is **Planck's constant**. The standard energy unit joules (J) is related to the nonstandard energy unit electron volts (eV) by 1 eV $= 1.602 \times 10^{-19}$ J.

The first experiment showing the photon nature of light demonstrated what is now known as the **photoelectric effect**. The apparatus for a photoelectric experiment is shown here. Light of a variable but known frequency f strikes a metal surface C in an evacuated tube, causing electrons to be emitted from C with various kinetic energies. After traversing a short distance in the tube, the electrons are collected by a second metal plate A. The electrons collected at the plate A constitute a current that is measured by the ammeter. In the photoelectric experiment, the current is measured as a function of the frequency and the intensity of the incident light. In addition, the maximum kinetic energy K_{max} of the electron is also measured.

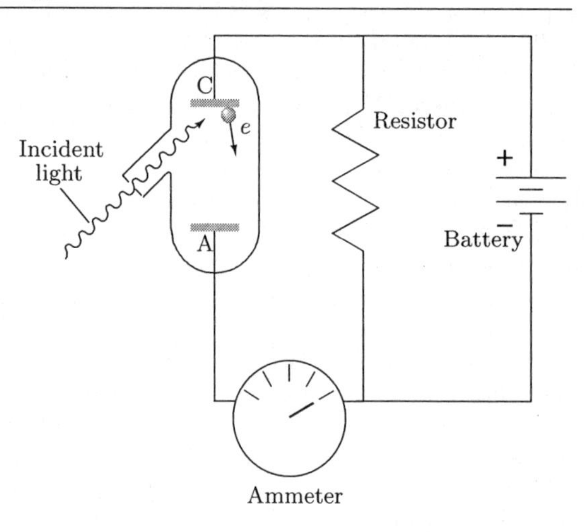

The photoelectric experiment has five main results:

1. Electrons emitted from the surface C have kinetic energies ranging from zero to a maximum value K_{max}.
2. K_{max} does not depend on the intensity of the incident light.
3. K_{max} does depend linearly on the frequency of the incident light according to a relationship known as **Einstein's photoelectric equation**: $K_{max} = \left(\frac{1}{2}mv^2\right)_{max} = hf - \phi$. The constant ϕ, called the **work function**, is equal to the energy needed to remove the least tightly bound electrons from the surface of C and is, therefore, a characteristic of the particular metal composing C.
4. Photons with frequencies below a smallest threshold frequency f_t (or with wavelengths longer than a longest threshold wavelength λ_t) do not have enough energy to eject electrons from the surface C, no matter how intense the incident light. This means that for f_t and below (or λ_t and above), $K_{max} = 0$, which allows us to calculate the threshold frequency and wavelength from Einstein's photoelectric equation.
5. The current is observed a very short time, of the order of 10^{-9} s, after the incident light strikes C, no matter how low the intensity of the light.

The classical picture of light composed of electromagnetic waves does not explain the results of the photoelectric experiment. Classically, the energy in a beam of light depends only on the light's intensity—the frequency of the incident light plays no role at all in energy balance. Classically the electrons in surface C would absorb more energy from more intense light and eventually, after a significant time lapse, leave C with kinetic energies ranging from zero to infinity.

Einstein explained all the experimental results of the photoelectric experiment by assuming that the light incident on the surface C is composed of photons, all with the same energy $E = hf$, where h is the same constant determined by Planck in his blackbody analysis about five years previously. Each photon interacts with an electron in the emitter C in an all-or-nothing fashion, either giving all its energy to an electron or not interacting at all.

Electrons leave C with varying kinetic energies because some lose energy as they travel through the material of C. An electron at the surface of C requires the smallest amount of energy—equal to the work function ϕ of the material—in order to be released. Such an electron is emitted with the maximum kinetic energy K_{max} given in Einstein's photoelectric equation, which is equal to the energy hf of the absorbed photon minus the energy ϕ required to release the least tightly bound electron from the surface of C. Thus Einstein's quantum picture of light explains how the maximum kinetic energy of an emitted electron depends on the light's frequency, why there is a threshold frequency f_{t} for which $K_{\text{max}} = 0$, and why the experimentally observed threshold frequency depends on the work function of the material through $hf_{\text{t}} = \phi$.

In the picture of light as photons, higher intensity corresponds to more photons in a light beam; and more photons eject more electrons from C. If the frequency of the incident photons stays constant, the maximum kinetic energy of the ejected electrons does not change as the intensity of the photons varies, which agrees with experimental results. The all-or-nothing transfer of energy from a photon to an electron also accounts for the exceedingly short time it takes for a current to be observed after light strikes C.

In classical theory, the energy and momentum of an electromagnetic wave are related by $E = pc$ or $p = E/c$. Applying the quantum relationship $E = hf$ together with the wave relationship $c = \lambda f$ gives the **momentum of a photon** as $p = hf/c = h/\lambda$.

In 1923, Arthur H. Compton provided additional evidence that the photon picture of light is correct by his relativistic analysis of a "billiard ball" type of experiment between a photon and an effectively free electron. In a **Compton experiment**, a photon with initial energy $E_{\text{i}} = hc/\lambda_{\text{i}}$ and momentum $p_{\text{i}} = h/\lambda_{\text{i}}$ is sent on a collision course with an electron that is at rest. After an elastic collision, the photon moves off at an angle θ with its original direction with a different energy $E_{\text{s}} = hc/\lambda_{\text{s}}$ and different momentum $p_{\text{s}} = h/\lambda_{\text{s}}$ and the electron moves at an angle ϕ with the original direction of the photon, as shown.

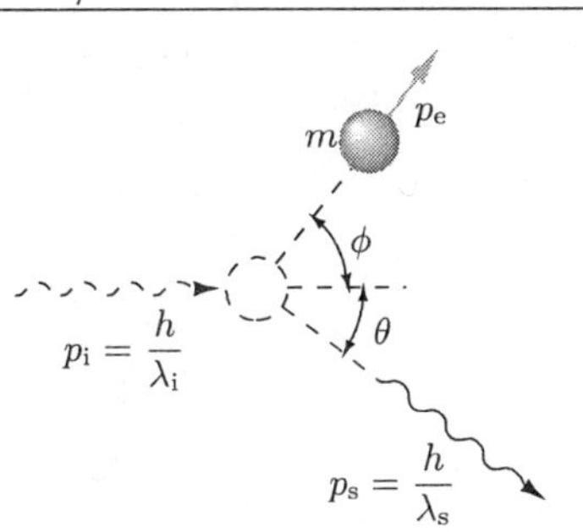

Compton applied relativistic conservation of energy and momentum to the collision (see Chapter R). The end result is that the wavelength λ_{s} of the photon scattered at angle θ after the collision is related to the wavelength λ_{i} of the incident photon before the collision by the **Compton equation** $\lambda_{\text{s}} - \lambda_{\text{i}} = [h/(m_{\text{e}}c)](1 - \cos\theta)$. The quantity $h/(m_{\text{e}}c) = 2.43$ pm $= \lambda_{\text{C}}$ is called the **Compton wavelength**. This short wavelength means that X rays or gamma rays are typically needed to observe a $\Delta\lambda$. For Compton scattering from a particle other than an electron, you must use the mass of that particle in the above expressions.

Physical Quantities and Their Units

Planck's constant	$h = 6.626 \times 10^{-34}\ \text{J}\cdot\text{s} = 4.136 \times 10^{-15}\ \text{eV}\cdot\text{s}$
Compton wavelength	$\lambda_C = 2.43\ \text{pm}$

Fundamental Equations

Photon energy	$E = hf = \dfrac{hc}{\lambda}$
Photon momentum	$p = \dfrac{h}{\lambda}$

Important Derived Results

Einstein's photoelectric equation	$K_{\text{max}} = \left(\dfrac{1}{2}mv^2\right)_{\text{max}} = hf - \phi$
Threshold frequency and wavelength	$\phi = hf_t = \dfrac{hc}{\lambda_t}$
Compton equation	$\lambda_s - \lambda_i = \dfrac{h}{m_e c}(1 - \cos\theta)$

Common Pitfalls

> Do not simply plug numbers into the photoelectric equation when working photoelectric effect problems. Understand the energy transfers that are going on.

1. TRUE or FALSE: The Compton wavelength of a proton is shorter than the Compton wavelength of an electron.

2. Explain why there is a threshold frequency in a photoelectric experiment.

Try It Yourself #1

In a photoelectric effect experiment, it is found that the maximum kinetic energy of emitted electrons for 400-nm light is 2.02 eV. Find the threshold wavelength for the emitter surface.

Picture: Use the energy balance of Einstein's photoelectric equation to solve for the work function of the material, which can be used to find the threshold wavelength.

Solve:

Write Einstein's photoelectric equation.	

Substitute the values given, with units, into the expression to solve for the work function.	
Relate the work function to the threshold wavelength and solve.	$\lambda_t = 1148$ nm

Check: The units work out properly. This wavelength is in the IR part of the spectrum, and so seems reasonable.

Taking It Further: If a material with a higher work function were to be used, would the threshold wavelength increase or decrease? Explain.

Try It Yourself #2

Find the wavelength of a 0.600-MeV photon after scattering at an angle of 70° in a Compton scattering experiment. What is the energy of the scattered electron?

Picture: Use the energy to find the incident wavelength. Then use the Compton relationship to find the scattered wavelength. The energy of the electron will be the difference between the incident and scattered photon energies.

Solve:

Determine the incident wavelength from the energy.	

Use the Compton relationship to find the scattered wavelength.	$\lambda_s = 3.66$ pm
Determine the energy of the scattered photon.	
The energy of the electron is the energy difference between the incident and scattered photons.	$E_e = 0.261$ MeV

Check: This energy is roughly half the rest energy of an electron, so seems reasonable.

Taking It Further: Describe how the energy of the scattered electron changes as the Compton angle is increased to 180°. Explain.

34.4 Energy Quantization in Atoms

In a Nutshell

If atoms in a low-pressure gas are excited by an electrical discharge, they emit light of specific wavelengths that are characteristic of the element or the compound. In 1913, this observation led Niels Bohr to postulate that the internal energy of an atom can have only a discrete set of values—that is, the internal energy of an atom is quantized. The Bohr model of the atom is still useful today, even though the reason for this quantization was not understood for another decade.

34.5 Electrons and Matter Waves

In a Nutshell

We have seen that a photon has the particle attributes of energy $E = hf$ and momentum $p = h/\lambda$. In addition, light shows wave behavior in its interference and diffraction. In 1924, after consulting Einstein, de Broglie boldly suggested in his doctoral dissertation that if photons behave both like waves and like particles, then perhaps electrons, which were regarded as "particles," could also have this dual character. de Broglie used the momentum expression for a photon to define the **wavelength of an electron** $\lambda = h/p$, where $p = mv$ is the (nonrelativistic) momentum of the electron. The expression for the frequency of the wave associated with an electron is the same as that for a photon: $f = E/h$, where E is the energy of the electron. Eventually, the wave nature of electrons, protons, and neutrons was substantiated experimentally.

The first demonstration of the wave nature of "particles" occurred in 1927 by C. J. Davisson and L. H. Germer at the Bell Telephone Laboratories. Davisson and Germer studied electron scattering from a nickel target. They found that the electron intensity in the scattered beam as a function of scattering angle showed maxima and minima corresponding to the exact de Broglie wavelength associated with the energy of the incident electron beam.

Also in 1927, G. P. Thomson (son of J. J. Thomson, who discovered the existence of the electron) demonstrated electron diffraction by sending electron beams through thin metal foils. The radii of the resulting diffraction pattern of concentric circles agreed with the de Broglie wavelength associated with the energy of the incident electron beam.

An important application of de Broglie waves associated with electrons is the electron microscope, which employs electrons to "see" objects at scales far smaller than microscopes using visible light. Also, diffraction is now routinely observed with "particles" other than electrons, such as neutrons.

When an object such as an electron is confined in a certain spatial region, standing waves occur for only certain wavelengths and frequencies consistent with the boundary conditions. This is similar to the effect of boundary conditions on standing waves in a string. As will be discussed in more detail in Chapter 35, Schrödinger and others showed, about 1928, that the application of boundary conditions to de Broglie waves led to a new fundamental description of nature called **quantum theory**, **quantum mechanics**, or **wave mechanics**, built around the concept of a wave function that satisfies an equation known as Schrödinger's equation. A consequence of quantum theory is that the energy of a bound system is quantized—that is, the energy of the system cannot be continuous, but can have only certain discrete values.

Fundamental Equations

de Broglie wavelength of a "particle"	$\lambda = h/p$

Common Pitfalls

> Energy is *not* always quantized. Quantization occurs only when particles are bound together—that is, one particle cannot classically "escape" another particle.

3. TRUE or FALSE: The de Broglie wavelength of a neutron is smaller than the de Broglie wavelength of an electron that has the same momentum.

4. An electron is accelerated from rest, acquiring a kinetic energy K. How does the de Broglie wavelength of the electron depend on K?

Try It Yourself #3

A typical kinetic energy of neutrons in a nuclear reactor is around 0.0400 eV. What is the de Broglie wavelength of such a neutron?

Picture: Determine the neutron's momentum, and use that to calculate its wavelength.

Solve:

Find an *algebraic* expression for the neutron's momentum from its kinetic energy and mass.	
Use the momentum to determine the de Broglie wavelength of the neutron.	$\lambda = 0.143$ nm

Check: The units work out properly.

Taking It Further: What must be the kinetic energy of an electron for it to have the same de Broglie wavelength as the neutron in this problem? Explain.

34.6 The Interpretation of the Wave Function

In a Nutshell

Classical physics is built on the notion that it is possible to determine the location of a particle such as an electron with unlimited precision; thus its trajectory—its location as a function of time—can be known exactly. Quantum mechanics paints an entirely different picture of the entity that was regarded classically as a "particle." According to quantum mechanics, we can determine only the *probability* of finding a particle, such as an electron, in a small space around a point. In one dimension, quantum mechanics specifies the probability of finding an electron somewhere in a given spatial interval dx, rather than exactly at some point x.

The probability of finding a particle in an interval is proportional to the size of the interval. If you increase the size of dx, the probability of finding the particle in the interval increases as well. There is never any certainty that you will find an electron in the interval—there is only a probability. The one thing that can be said with certainty is that you will never find an electron in an interval where the probability is zero.

In a probability description, the **probability density** $P(x)$ gives the likelihood of finding a particle in the vicinity of one value of x rather than another value. The probability of finding a particle in a spatial interval from x to $x + dx$ is $P(x)\,dx$. In the quantum mechanical picture, the quantity that determines the probability density $P(x)$ is called the **wave function** $\psi(x)$. $P(x)$ and $\psi(x)$ are related by $P(x) = \psi^2(x)$. Thus, $\psi^2(x)\,dx$ is the probability of finding a particle in a spatial interval from x to $x + dx$: probability $= \psi^2(x)\,dx$.

The functional form of the wave function $\psi(x)$ is determined from the Schrödinger wave equation, which will be described in Chapter 35. You can think of $\psi(x)$ as the entity that exhibits the wavelike properties of de Broglie waves, which result in the wave properties of objects.

The probability of finding an object in the interval dx is $\psi^2(x)\,dx$, which represents the probability of finding an object in the spatial interval between x and $x + dx$. The object must certainly be somewhere between $x = -\infty$ and $x = +\infty$, so the sum of the probabilities over all intervals must equal 1: $\int_{-\infty}^{\infty} \psi^2(x)\,dx = 1$. This defines a **normalization condition** that the wave function $\psi(x)$ must satisfy. In addition, the wave function must approach zero as x approaches plus or minus infinity.

Physical Quantities and Their Units

Wave function	$\psi(x)$

Important Derived Results

Probability density	$P(x) = \psi^2(x)$
Probability of finding an object in a spatial interval from x to $x + dx$	Probability $= P(x)\,dx = \psi^2(x)\,dx$
Normalization condition	$\int_{-\infty}^{\infty} \psi^2(x)\,dx = 1$

Common Pitfalls

> It can be easy to confuse the probability density with probability. The probability density is a function defined at each point x. Probability refers to the probability of finding an object in a small spatial interval dx that straddles a point x; for example, the probability of finding an object in an interval between x and $x+dx$. The probability density is related to the probability by $P(x)\,dx =$ probability.

> Understand that the probability density $P(x)$ is related to the square of the wave function $\psi(x)$ by $P(x) = \psi^2(x)$.

5. TRUE or FALSE: The probability of finding an object in a given interval dx is proportional to the value of the wave function at the location of dx.

6. What two quantities determine the probability of locating an object?

Try It Yourself #4

The ground-state wave function for a quantum mechanical harmonic oscillator (see Section 34-10) is $\psi(x) = C\,e^{-x^2/(2A^2)}$, where C and A are constants for a given harmonic oscillator. Find how C is related to A so that this wave function is normalized.

Picture: Use the normalization condition. You will have to use an integral table to evaluate the integral.

Solve:

Write the normalization integral.	
Use an integral table to evaluate the integral.	
Solve for C in terms of A.	$C = A^{-1/2}\pi^{-1/4}$

Check: With this substitution, the normalization integral evaluates to 1.

Try It Yourself #5

The wave function for the second excited state of a particle in a box of length L (see Section 34.8) is given by $\psi(x) = \sqrt{2/L}\sin(3\pi x/L)$. Find the probability of finding the particle in this excited state in the central region $L/4 < x < 3L/4$ of the box.

Picture: Determine the probability distribution from the wave function and integrate over the region of interest.

Solve:

Write an expression for the probability from the wave function, but do not evaluate the integral.	
Evaluate the indefinite integral using a standard integration table.	
Solve for the probability by applying the appropriate limits.	probability = 0.394

Check: The probability is less than 1, which is good. This result tells us that there is about a 39 percent chance of finding the particle in the central region of the box. As a result it spends more time in the turnaround zones.

Taking It Further: Is this wave function normalized? How can you tell?

34.7 Wave–Particle Duality

In a Nutshell

At times objects exhibit wave properties and at other times particle properties. This phenomenon is known as **wave–particle duality**. Both the wave and particle pictures are necessary for a complete understanding of an object, but you cannot observe both aspects simultaneously in a single experiment, which is the **principle of complementarity**, first put forth by Bohr in 1928.

If Young's two-slit experiment is repeated with a beam of electrons instead of visible light, a similar interference pattern of bright and dark bands of electrons is observed on a screen or film. The separation between the bright and dark bands depends upon the de Broglie wavelength associated with the energy of the electron beam. One can watch the buildup of the bands as each individual electron strikes the screen. The buildup is determined by the probability of detecting an electron on the screen, which is proportional to $\psi^2(x)$.

In a thought experiment, when we say we have located a particle at a given position x, we really mean we have determined that its position is somewhere in the interval between $x - \Delta x$ and $x + \Delta x$, where Δx is the uncertainty in the position x. Similarly, if we measure the momentum p of a moving object, we really mean that we have determined that the object's momentum is somewhere in the interval between $p - \Delta p$ and $p + \Delta p$, where Δp is the uncertainty in the momentum p.

In classical physics, Δx and Δp can in principle be individually made to be arbitrarily small. In fact, a basic premise in describing the motion of a classical particle is that rigorously $\Delta x = 0$ and $\Delta p = 0$ so that one can talk unambiguously about the exact instantaneous position of a particle along with its exact instantaneous velocity or momentum.

When uncertainties are analyzed taking quantum considerations into account, one finds that the position and momentum of an object cannot be specified with infinite precision. Suppose, for example, you want to determine the position of an electron by "looking" at it. To do this measurement, you must bounce a photon off the electron. The interaction of the photon with the electron as the bouncing takes place causes the electron's velocity and momentum to change by an unknown amount.

A rigorous treatment of the quantum-mechanical measurement process, first put forth by Werner Heisenberg in 1927, shows that in a simultaneous determination of the position and momentum of a particle such as an electron, the uncertainty in the momentum, Δp, and position, Δx, are related by the uncertainty principle: $\Delta x \, \Delta p \geq \hbar/2$, where $\hbar$ (read "h bar") $= h/(2\pi)$. The equality gives intrinsic lower limits, and represents the best uncertainty that is possible. If you try to locate the position of a particle more and more precisely by making Δx smaller and smaller, the uncertainty Δp in the particle's momentum becomes larger and larger, and vice versa. In any actual experiment, additional experimental uncertainties produce values of Δp and Δx that are larger than the intrinsic lower limits resulting from wave–particle duality, giving rise to the inequality in the uncertainty relationship.

Physical Quantities and Their Units

"h bar" $\hbar = \dfrac{h}{2\pi} = 1.05 \times 10^{-34}\ \mathrm{J \cdot s} = 0.658 \times 10^{-15}\ \mathrm{eV \cdot s}$

Fundamental Equations

Heisenberg's uncertainty principle for position and momentum $\Delta p \, \Delta x \geq \dfrac{1}{2}\hbar$

Common Pitfalls

> Heisenberg's uncertainty principle gives only a lower limit to the product $\Delta x \, \Delta p$. There is no upper limit.

7. TRUE or FALSE: According to the uncertainty principle, the more uncertain a particle's momentum, the more uncertain its position.

8. Describe an experiment that measures the wave and particle aspects of an object simultaneously.

34.8 A Particle in a Box

In a Nutshell

A "particle in a box" refers to a particle of mass m that is confined to a one-dimensional region of length L and can never be found outside the boundaries of the box. Very importantly, the energies of a particle in a box are quantized—only certain discrete values are possible. Although the concept of a "particle in a box" is somewhat artificial and does not truly exist in nature, such a particle exhibits many properties, such as discrete energies, similar to those of an electron bound within an atom, or a proton inside a nucleus.

Let the box be between $x = 0$ and $x = L$. Since the particle is confined to the box, it can never be found outside the boundaries of the box. This means that the wave function ψ is zero outside the boundaries of the box. Wave functions must be continuous since they represent probability amplitudes. A consequence of the **boundary conditions** that $\psi = 0$ at $x = 0$ and $x = L$, required by the continuity of the wave function, is that the energy of the particle cannot be continuous, as in classical theory. Rather, the particle's energy can take on only one of the quantized (discrete) values given by $E_n = n^2h^2/(8mL^2) = n^2E_1$, where the lowest or **ground-state energy** is given by $E_1 = h^2/(8mL^2)$. The integers $n = 1, 2, 3, \ldots$ are called **quantum numbers**.

The smallest possible value of the energy is not zero, as in classical theory. Rather, the smallest (ground state) energy, called the **zero-point energy**, is given by the expression for E_1. This means that a particle in a box can never be at rest but must always be moving with one of the allowed energies given by E_n. Since E_1 varies as $1/L^2$, the smaller the confines of the box, the larger the zero-point energy. We show here the first five energy levels of a particle in a box.

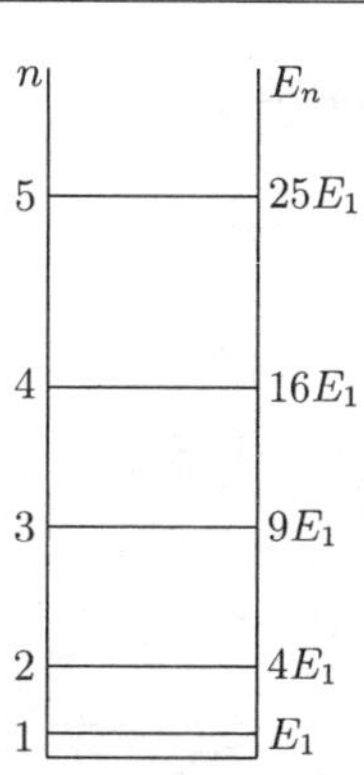

The relation of quantized energies to the size of a box is an expression of the uncertainty principle. We know the particle is inside the box, but we do not know exactly where inside the box the particle is, so the uncertainty in the particle's position is $\Delta x = L$. The particle's energy can be any of the quantized values $E_n = n^2E_1$. A measure of the uncertainty in the particle's energy can be taken as $\Delta E = E_1$. The relationship between a particle's momentum and energy is $p = \sqrt{2mE}$, so the uncertainty Δp in momentum corresponding to the uncertainty in energy is $\Delta p = \sqrt{2m\,\Delta E} = \sqrt{2mE_1} = \sqrt{2mh^2/(8m\,(\Delta x)^2)} = h/(2\,\Delta x)$, from which $\Delta x\,\Delta p = h/2$, which is a statement of the Heisenberg uncertainty principle for a particle in a box.

It is possible for a system that has quantized energies, such as a particle in a box, to make a transition from an initial energy state E_i to a different final energy state E_f. If $E_\mathrm{i} > E_\mathrm{f}$ a photon will be emitted; if $E_\mathrm{i} < E_\mathrm{f}$ a photon will be absorbed. From conservation of energy, the energy of the emitted or absorbed photon is equal to the energy difference between the initial and final states of the system, from which the frequency and wavelength of the emitted radiation can be found.

The wave functions ψ_n obtained by solving the Schrödinger equation for a particle in a box between $x = 0$ and $x = L$ satisfying the normalization condition, which will be done in Chapter 35, have the same form as for a vibrating string: $\psi_n = \sqrt{2/L}\sin(n\pi x/L)$, where $n = 1, 2, 3, \ldots$. The corresponding probability distribution is $P_n = \psi_n^2 = (2/L)\sin^2(n\pi x/L)$. Plots of $\psi_n(x)$ and $\psi_n^2(x)$ are shown for the ground state ($n = 1$) and the first two excited states ($n = 2, 3$). For each standing wave function, the particle has an energy E_n as given above.

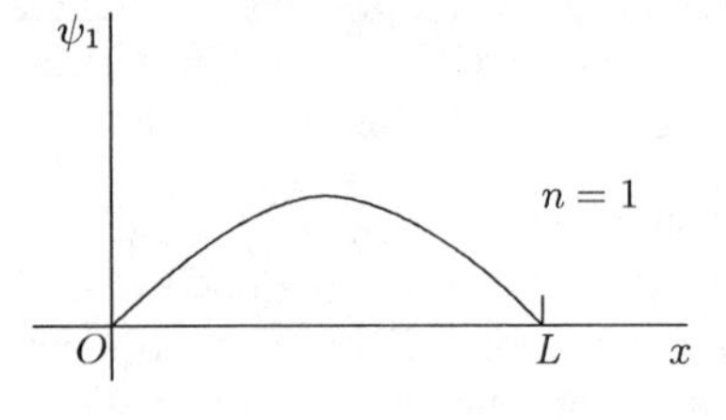

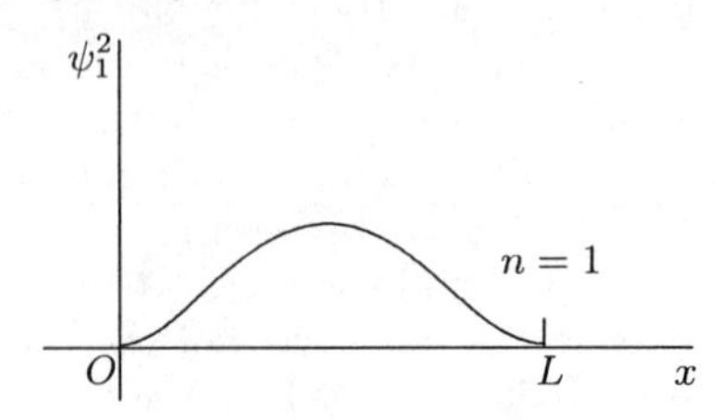

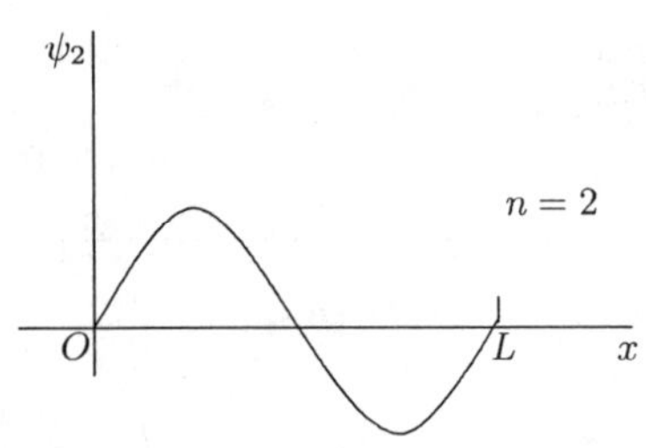

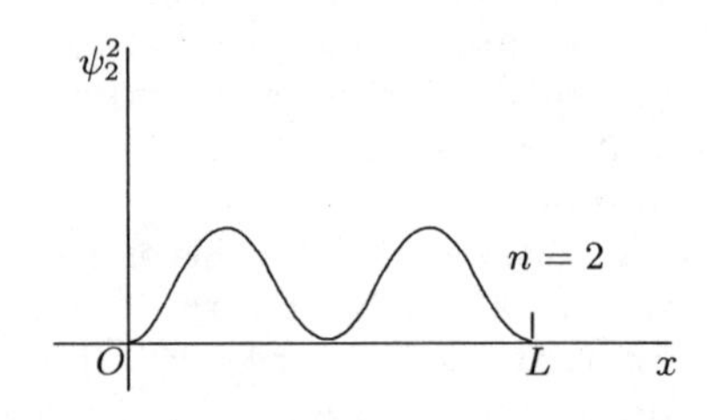

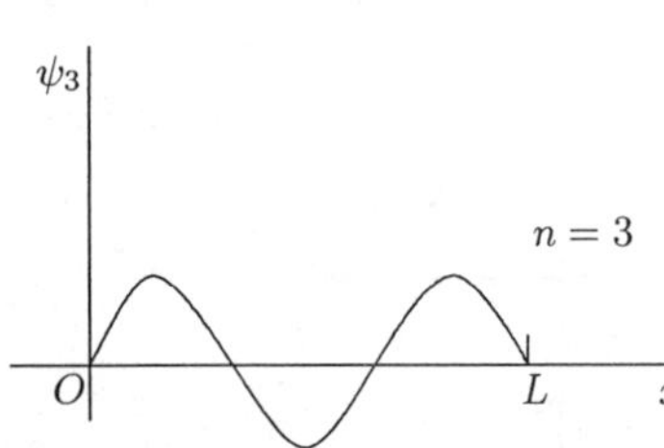

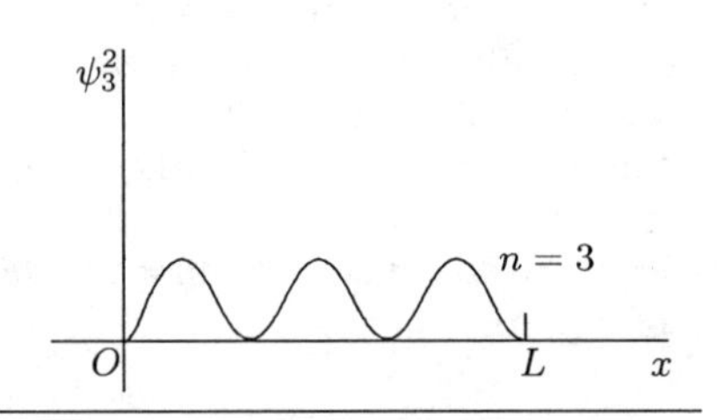

For very large values of the quantum number n, the maxima of the probability distribution $P_n\psi_n^2$ are so closely spaced that ψ^2 cannot be distinguished from its average value. The fact that the probability distribution is nearly constant across the whole box means that the particle is nearly equally likely to be found anywhere in the box, which is the same as the classical result. An example of this is shown for $n = 10$. This is an example of **Bohr's correspondence principle**: "In the limit of very large quantum numbers, the classical calculation and the quantum calculation must yield the same results."

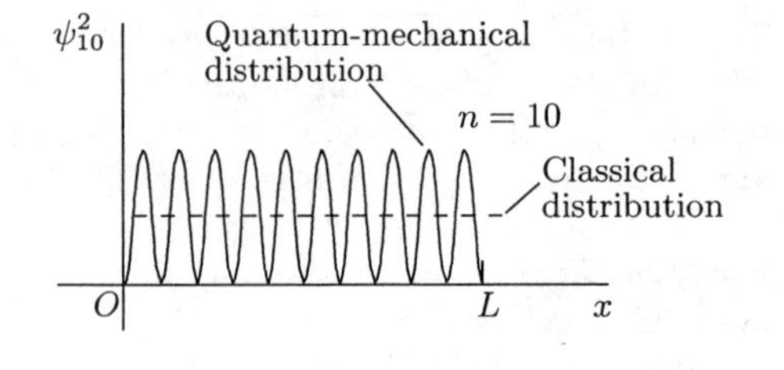

Important Derived Results

Allowed energies for a particle in a box	$E_n = n^2 \dfrac{h^2}{8mL^2} = n^2 E_1$, for $n = 1, 2, 3, \ldots$
Ground-state energy for a particle in a box	$E_1 = \dfrac{h^2}{8mL^2}$
Wave function for a particle in a box lying between $x = 0$ and $x = L$	$\psi_n = \sqrt{\dfrac{2}{L}} \sin\left(\dfrac{n\pi x}{L}\right)$ for $n = 1, 2, 3, \ldots$

Common Pitfalls

> You might think that there is something unphysical about a wave function. However, according to quantum mechanics, a wave function is as physical a description of an object (such as an electron) as we can get. After all, how real is a point particle in classical physics?

9. TRUE or FALSE: When a particle is confined to a certain region of space in a system, the energy of an emitted photon equals the energy difference between the quantized initial and final energy levels of the system through which the particle makes a transition.

10. Describe what happens to the energy levels of a particle in a box as the length of the box increases.

Try It Yourself #6

A particle in a one-dimensional box of length L is in its first excited state. If you start from $x = 0$, at what value of x will there be a 25 percent probability of finding the particle in this interval?

Picture: Using the wave function for the first excited state, determine the probability distribution function and integrate to find the probability. Keeping in mind that the probability equals 1 for finding the particle in the interval from $x = 0$ to $x = L$, draw a picture of the wave function and the probability to help you guess a simple solution to a complicated transcendental equation.

Solve:

Write the general *algebraic* expression for the wave functions of a particle in a box.	
Write the *algebraic* expression for the first excited state, which is the second state overall.	

Sketch the wave function over the range of $0 \leq x \leq L$.	
The probability is the square of the wave function. Sketch the square of the wave function over the range of $0 \leq x \leq L$. Remember, the total integrated probability is 1.	
From your sketch of the probability, it should be relatively easy to determine the value of x for which the probability of finding the particle in the range from 0 to x is 25 percent.	$x = L/4$

Check: The value of x certainly must be between 0 and L, which it is.

Taking It Further: How does the answer change, if at all, if we use the third excited state? Explain.

Try It Yourself #7

The size of a nucleus is about 10^{-14} m. Treating the nucleus as a one-dimensional particle in a box, find the ground-state energy of a proton and an electron if each particle were bound in the nucleus.

Picture: Substitute the given values into the ground-state energy expression.

Solve:

Write the *algebraic* expression for the ground state energy of a particle in a box.	
Substitute the given values for a proton.	$E_{\text{proton}} = 2.05$ MeV
Substitute the given values for an electron.	$E_{\text{electron}} = 3760$ MeV

Check: This energy is reasonable for the proton. However, the ground-state energy for an electron is extremely high, equivalent to the kinetic energy of an electron traveling at $v = 0.99993c$.

Taking It Further: Given that the potential energy well is so deep (it can capture electrons with the high speed given above), what happens to electrons with less energy? Explain.

34.9 Expectation Values

In a Nutshell

A typical problem in classical mechanics is the specification of the exact position of a particle as a function of time. In contrast, according to quantum mechanics it is intrinsically impossible to specify exactly the position of a particle at any given time. The most we can know about the position of a particle is the relative probability of finding it in one interval or another. However, if you measure the position x of a particle in a large number of identical experiments, you will obtain some average value of x, designated by $\langle x\rangle$. This average value is called the **expectation value** and is related to the wave function ψ by $\langle x\rangle = \int_{-\infty}^{\infty} x\psi^2(x)\,dx$. If, instead of the position x, you look at the expectation value of any function $F(x)$, you observe $\langle F(x)\rangle = \int_{-\infty}^{\infty} F(x)\psi^2(x)\,dx$.

Probabilities and Expectations Problem-Solving Strategy

Solve:

1. To calculate the probability P of finding a particle in the region of infinitesimal length between x and $x + dx$, we multiply the length dx by the probability per unit length at x, where the probability per unit length (called the probability density function) is given by ψ^2.
2. To calculate the probability P of finding a particle in the region $x_1 < x < x_2$, we, in principle, divide the region into an infinite number of regions of infinitesimal length dx, calculate the probability P of finding the particle in each infinitesimal length, and then sum the probabilities—that is, we evaluate the integral $\int_{x_1}^{x_2} \psi^2\, dx$.
3. To calculate the expected value of a function $F(x)$, we evaluate the integral $\int_{-\infty}^{\infty} F(x)\psi^2(x)\, dx$. The result of this calculation is called the expected value of $F(x)$.

Important Derived Results

Expectation value of a function $$\langle F(x)\rangle = \int_{-\infty}^{\infty} F(x)\psi^2(x)\, dx$$

Try It Yourself #8

The ground-state wave function for a quantum mechanical harmonic oscillator is $\psi(x) = C\, e^{-x^2/(2A^2)}$, where C and A are constants for a given harmonic oscillator. Find the expectation value of x^2.

Picture: Substitute the given functions into the expression for the expectation value.

Solve:

Write the general *algebraic* expression for the expectation value of a function $F(x)$.	
Substitute the given functions into the expectation value expression.	
Determine how to calculate this integral from a standard integration table, by looking for the integral from 0 to ∞ instead of from $-\infty$ to ∞.	

Because this integral is symmetric, you should now easily be able to find the expectation value of x^2.	$\langle x^2 \rangle = \frac{1}{2} C^2 A^3 \sqrt{\pi}$

Check: It is interesting to note that the expected value of x^2 is not zero.

Taking It Further: What is the expectation value of x?

34.10 Energy Quantization in Other Systems

In a Nutshell

When you learn how to solve the Schrödinger equation for quantum mechanical systems in Chapter 35, a key component will be the **potential energy function** $U(x)$ of the system you are looking at. For a particle in a box located between $x = 0$ and $x = L$, which we have looked at above, the potential energy function $U(x)$ is given by

$$U(x) = \begin{cases} 0 & 0 < x < L \\ \infty & x < 0 \text{ or } x > L \end{cases}$$

The interpretation of the potential energy function $U(x)$ is that the particle can be found only between the coordinate values $x = 0$ and $x = L$. The particle will never be found outside the box in the region $x < 0$ or $x > L$.

The **harmonic oscillator** is a physically more realistic system than a particle in a box. With a harmonic oscillator, a particle of mass m is attached to a spring of force constant k and undergoes small oscillations of amplitude A about a fixed equilibrium point located at $x = 0$. The classical potential energy function $U(x)$ for a harmonic oscillator is $U = \frac{1}{2}kx^2 = \frac{1}{2}m\omega_0^2 x^2$, where $\omega_0 = \sqrt{k/m}$ is the natural angular frequency in rad/s of the harmonic oscillator. The natural oscillation frequency f_0 in vibrations per second is related to ω_0 by $\omega_0 = 2\pi f_0$, so $f_0 = \sqrt{k/m}/(2\pi)$. Classically, the object oscillates between $x = +A$ and $x = -A$.

The allowed energies of a particle in a quantum-mechanical box are quantized. Similarly, a quantum-mechanical analysis of a harmonic oscillator particle (see Chapter 35) shows that the allowed energies are quantized, with the discrete energies E_n given by $E_n = (n + \frac{1}{2})hf_0$, where $n = 0, 1, 2, \ldots$, and the **ground-state energy** (lowest energy level) is $E_0 = \frac{1}{2}hf_0$. The allowed energy levels are shown, along with the potential energy function $U(x)$. Note that the energy levels for a harmonic oscillator are evenly spaced, as compared to the unevenly spaced energy levels for a particle in a box.

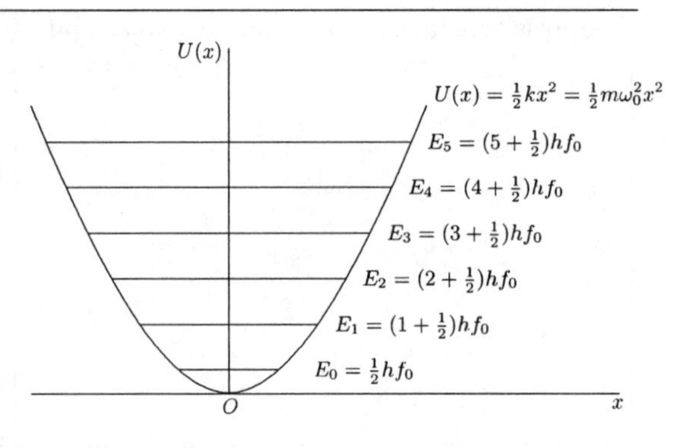

One main result of a quantum analysis of a particle bound in a box or bound in a harmonic oscillator is that only discrete energies are allowed. Energies are quantized and are described by quantum numbers n. A similar situation arises in a **hydrogen atom**, where an electron is bound to a proton by the electrostatic force of attraction of a nucleus. As will be shown in Chapter 36, the allowed energies of the electron in a hydrogen atom are given by the quantized values $E_n = -13.6 \text{ eV}/n^2$, for $n = 1, 2, 3, \ldots$. The lowest energy, corresponding to $n = 1$, is the ground-state energy $E_1 = -13.6$ eV.

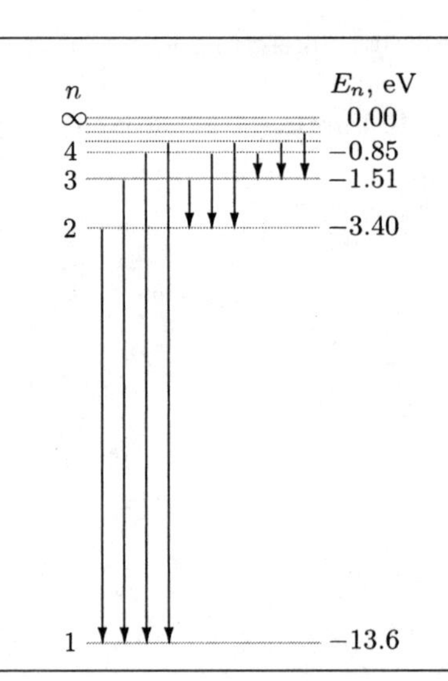

Physical Quantities and Their Units

Ground state energy of the hydrogen atom	$E_1 = -13.6$ eV

Important Derived Results

Allowed energies for a quantum-mechanical harmonic oscillator	$E_n = \left(n + \frac{1}{2}\right) hf_0$, for $n = 0, 1, 2, 3, \ldots$
Allowed energies of an electron in a hydrogen atom	$E_n = -\dfrac{13.6 \text{ eV}}{n^2}$, for $n = 1, 2, 3, \ldots$

Common Pitfalls

> A free electron by itself is not quantized—it can have any arbitrary kinetic energy. The quantization of the energy of an electron occurs only when the electron is confined in some way (say as a particle in a box). When confined, quantization of energy arises from boundary conditions, which can occur only if there are boundaries. The three types of confinement discussed in this chapter are a particle in a box, a harmonic oscillator, and a hydrogen atom.

- ⪢ Be sure you understand the differences in the energy-level structures of a particle in a box, a harmonic oscillator, and a hydrogen atom. In particular, make sure you understand how the energy levels depend on the quantum number n for each system.
- ⪢ One main result of quantized systems such as a particle in a box, a harmonic oscillator, and the hydrogen atom is that the energies of the particle in the system are quantized. Make sure you understand that the energy $E = hf$ of an emitted or absorbed photon equals the difference of the energies of the energy levels between which the particle makes a transition.

11. TRUE or FALSE: In a quantum-mechanical harmonic oscillator, the frequency of a photon emitted in a transition from $n = 3$ to $n = 2$ is equal to the frequency of a photon emitted in a transition from $n = 2$ to $n = 1$.

12. Describe how energy levels depend upon the quantum number n for (a) a particle in a box, (b) a harmonic oscillator, and (c) a hydrogen atom.

Try It Yourself #9

For a harmonic oscillator, find the ratio of the frequency of a photon emitted from an $n = 3$–to–$n = 1$ transition to the frequency of a photon emitted from an $n = 2$–to–$n = 1$ transition.

Picture: The frequency can be found from the energy of the photon, which is equal to the energy difference of the states involved in the transition.

Solve:

Write the general *algebraic* expression for the energy states of a harmonic oscillator.	
Find the energy difference between the $n = 3$ and $n = 1$ states.	
Find the energy difference between the $n = 2$ and $n = 1$ states.	

The energy of a photon is given by hf, so we can easily determine the frequency of the two photons and their ratio.	$f_{31}/f_{21} = 2$

Check: Because the energies of a quantum-mechanical harmonic oscillator are equally spaced, the ratio should be an integer. The "harmony" in a harmonic oscillator arises from the fact that the frequency difference between each state is f_0.

Taking It Further: Will the ratio of frequencies for the same two transitions of the hydrogen atom be an integer? Explain.

Try It Yourself #10

A 2.00-kg object is attached to the end of a spring with a force constant of 400 N/m and oscillates without friction with an amplitude of 8.00 cm. If the energy of this classical harmonic oscillator were quantized according to quantum-mechanical rules, what is the value of the corresponding quantum number?

Picture: Equate the classical expression for the energy of a harmonic oscillator to the quantum mechanical expression for a harmonic oscillator and solve for n.

Solve:

Find the energy of this classical simple harmonic oscillator.	
Determine the natural frequency of this oscillator.	

Write an *algebraic* expression for the energy of a quantum mechanical harmonic oscillator.	
Equate the expressions from steps 1 and 2 to solve for n.	$n \approx 9 \times 10^{32}$

Check: Because the problem is macroscopic in nature, we should expect very large quantum numbers. This enormous number is an illustration of Bohr's correspondence principle, by which in the limit of very large quantum numbers, classical and quantum calculations must yield the same result.

QUIZ

1. TRUE or FALSE: Quantized energy levels arise because of confinement of a particle.

2. TRUE or FALSE: When a particle is confined to a certain region of space in a system, the energy of an emitted photon equals the energy difference between the quantized initial and final energy levels of the system through which the particle makes a transition.

3. An electron is accelerated from rest, acquiring a kinetic energy K. How does the de Broglie wavelength of the electron depend on K?

   Because $\lambda_c = h/p$, where $p = mv$ ∴ p & K are directly related.

4. In a photoelectric experiment, what effect does changing the material of the emitting surface C to a new material with a higher work function have on the emitted electrons?

   The emitted electrons require a higher energy interaction w/ a photon in order to be released. The new required E must at least equal the new ϕ.

5. In problem 7 of this quiz, you will find two apparently contradictory results for a particle in the first excited state of a box of length L. The first is that the probability density is zero for $x = L/2$, which means the particle will never be found in the middle of the box. The second is that the expectation value of the position $\langle x \rangle = L/2$. How can the expectation value of the position of the particle be located at a point where the particle will never be found?

   prob. density = $\int \psi^2 x^2 \, dx$

   expec. val. = $\int F(x) \psi^2(x) \, dx$ → ave. value

   ∴ expec. value ≠ exact location ∴ particle can symmetrically spend time around place where $\langle x^2 \rangle = 0$

6. The work function for potassium is 2.21 eV. In a photoelectric effect experiment using a potassium emitter, the maximum kinetic energy of the emitted electrons is observed to be 2.82 eV. What is the wavelength of the incident light?

7. Demonstrate that the probability density for a particle in the first excited state of a box of length L is zero at the center of the box. Also, find the expectation value $\langle x \rangle$ for the position of a particle in the first excited state of a box of length L.

Chapter 35

Applications of the Schrödinger Equation

35.1 The Schrödinger Equation

In a Nutshell

In general, the **one-dimensional wave function**, which we write as $\Psi(x,t)$, is a function of both the time coordinate t and the one-dimensional position coordinate x. The fundamental equation, the wave equation known as Schrödinger's equation, that gives $\Psi(x,t)$, cannot be derived, just as Newton's second law $\vec{a} = \sum \vec{F}/m$, relating the acceleration of a mass to the forces applied to the mass, cannot be derived. The **time-dependent Schrödinger equation** is

$$-\frac{\hbar^2}{2m}\frac{\partial^2\Psi(x,t)}{\partial x^2} + U\Psi(x,t) = i\hbar\frac{\partial\Psi(x,t)}{\partial t}$$

In this expression U is the potential energy function for the object, which could be the Coulomb potential energy, the potential energy of a spring, or (prevalent in the following discussions) a constant potential energy that changes its value in sudden jumps at certain locations. It should be noted that the Schrödinger equation involves the imaginary number $i = \sqrt{-1}$.

Because the wave function can in general be complex, we need to modify the probability density function introduced in Chapter 34, used to determine the probability of finding a particle in a region dx. $P(x,t)\,dx = |\Psi(x,t)|^2\,dx = \Psi^\star\Psi\,dx$, where $\Psi^\star$ is the **complex conjugate** of Ψ. The complex conjugate is identical to Ψ, except that i is replaced by $-i$ everywhere it appears in the expression for Ψ.

The standing-wave solution to the Schrödinger equation can be written as $\Psi(x,t) = \psi(x)\,\mathrm{e}^{-i\omega t}$, where we have separated the spatial and time dependence into two separate functions. In addition, $\mathrm{e}^{-i\omega t} = \cos(\omega t) - i\sin(\omega t)$.

Substituting the standing-wave solution into Schrödinger's equation and canceling the common $\mathrm{e}^{-i\omega t}$ term yields the **time-independent Schrödinger** equation:

$$-\frac{\hbar^2}{2m}\frac{d^2\psi(x)}{dx^2} + U(x)\psi(x) = E\psi(x)$$

where $E = \hbar\omega$ is the energy of the particle. In this book we will be concerned only with the time-independent Schrödinger equation.

One of the easiest problems to solve with the Schrödinger equation is a situation in which a particle is confined in a one-dimensional box of length L, which we discussed in Chapter 34. Because the electron can never possess sufficient energy to escape the box, this corresponds to an infinite square-well potential energy of the mathematical form

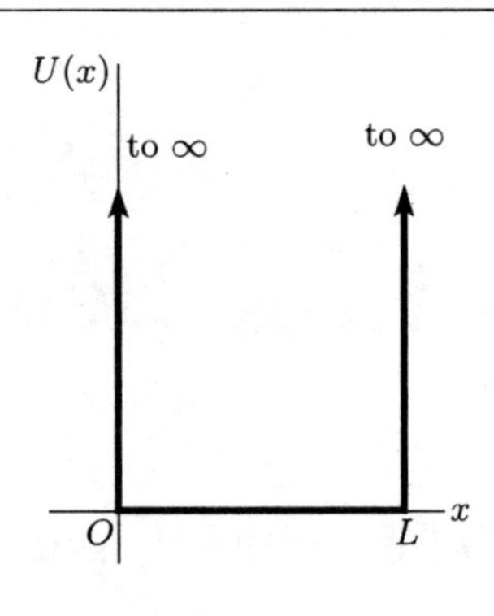

$$U(x) = \begin{cases} \infty & x < 0 \\ 0 & 0 < x < L \\ \infty & x > L \end{cases}$$

which is shown at the right.

The general solution to the time-independent Schrödinger equation inside the box is given by $\psi(x) = A \sin kx + B \cos kx$, where A and B are constants and $k = \sqrt{2mE}/\hbar$. The boundary condition that $\psi(x) = 0$ at $x = 0$ results in $B = 0$. The boundary condition that $\psi(x) = 0$ at $x = L$ requires that k take on discrete values given by $k_n = n\pi/L$, where $n = 1, 2, 3, \ldots$. Normalization requires that $A = \sqrt{2/L}$. The full solution to the time-independent Schrödinger equation is then given by

$$\psi(x) = \begin{cases} 0 & x < 0 \\ \sqrt{\dfrac{2}{L}} \sin \dfrac{n\pi x}{L} & 0 < x < L \\ 0 & x > L \end{cases}$$

The resulting energy levels are those presented in Chapter 34: $E_n = n^2h^2/(8mL^2) = n^2E_1$, where the ground-state energy is $E_1 = h^2/(8mL^2)$.

Fundamental Equations

Schrödinger's time-dependent equation
$$-\frac{\hbar^2}{2m}\frac{\partial^2\Psi(x,t)}{\partial x^2} + U\Psi(x,t) = i\hbar\frac{\partial\Psi(x,t)}{\partial t}$$

Schrödinger's time-independent equation
$$-\frac{\hbar^2}{2m}\frac{d^2\psi(x)}{dx^2} + U(x)\psi(x) = E\psi(x)$$

Important Derived Results

General solution to the time-dependent Schrödinger equation	$\Psi(x,t) = \psi(x)\,e^{-i\omega t}$
Infinite square-well potential energy	$U(x) = \begin{cases} \infty & x < 0 \\ 0 & 0 < x < L \\ \infty & x > L \end{cases}$
Wave function for the infinite square-well potential	$\psi(x) = \begin{cases} 0 & x < 0 \\ \sqrt{\frac{2}{L}} \sin \frac{n\pi x}{L} & 0 < x < L \\ 0 & x > L \end{cases}$
Ground-state energy for a particle in a box	$E_1 = \dfrac{h^2}{8mL^2}$
Allowed energies for a particle in a box	$E_n = n^2 E_1$ for $n = 1, 2, 3, \ldots$

Common Pitfalls

- It can be easy to confuse the time-dependent wave function $\Psi(x,t)$ with the time-independent wave function $\psi(x)$. The two are related by $\Psi(x,t) = \psi(x)\,e^{-i\omega t}$. You will usually be interested in the time-independent wave function $\psi(x)$.
- Remember that the wave function $\psi(x)$ is related to the probability density by $P = |\psi(x)|^2 = \psi^*(x)\psi(x)$.

1. TRUE or FALSE: Specification of a potential energy function in the time-independent Schrödinger equation determines a single unique wave function for a particle.

2. Describe the allowed wave functions for a particle in a box (infinite square-well potential).

Try It Yourself #1

Show that $\psi_n = A_n \sin(n\pi x/L)$ is a solution of the time-independent Schrödinger equation for a particle in a box of length L.

Picture: Calculate the first and second derivatives of ψ_n and substitute into Schrödinger's equation. If you end up with an energy that satisfies the quantization condition given in the text, then the provided wave function is a solution to Schrödinger's equation.

Solve:

Write the time-independent Schrödinger equation as a guide for the problem.	
Determine the potential energy function inside the box.	
Find the second derivative of $\psi(x)$ with respect to x.	
Substitute these expressions into Schrödinger's equation and solve for E.	$E = \dfrac{h^2n^2}{8mL^2}$

Check: The energy arrived at satisfies the quantization condition given in the text, so the given wave function must be a solution to Schrödinger's equation for a particle in a box.

35.2 A Particle in a Finite Square Well

In a Nutshell

For a particle in a square well whose walls have a finite (rather than an infinite) height, the potential energy that we put into the Schrödinger equation has the form

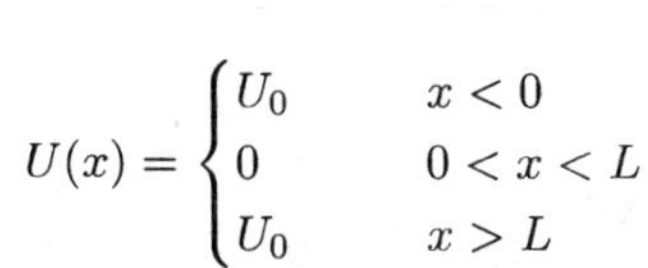

$$U(x) = \begin{cases} U_0 & x < 0 \\ 0 & 0 < x < L \\ U_0 & x > L \end{cases}$$

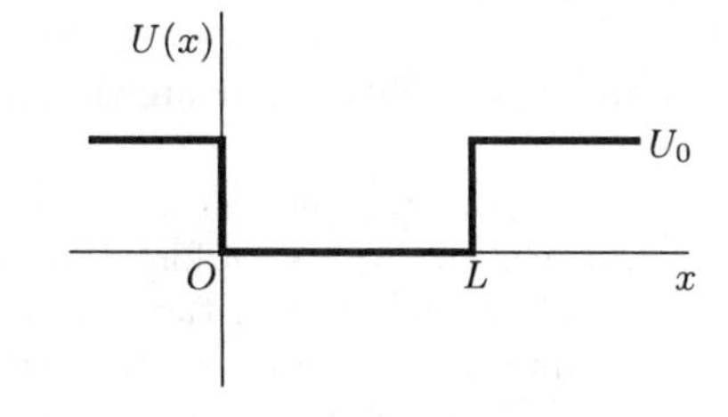

which is shown. When considering solutions to the Schrödinger equation for this potential function, we will consider only the case in which $E < U_0$.

Solving the Schrödinger equation with the finite square-well potential energy can be tricky, and matching the boundary conditions at $x = 0$ and $x = L$ is mathematically challenging and somewhat complicated. This is because the wave function is not zero outside the well. Consequently, we will not give mathematical expressions for ψ_n. Instead, we show the end results with graphs of the wave function ψ_n and the corresponding probability distribution $P_n = |\psi|^2$ for the ground state $n = 1$ and for the first two excited states ($n = 2$ and $n = 3$). These graphs are qualitatively similar to those for an infinite well, except the allowed wavelengths are somewhat larger and there are exponential "tails" that extend into the regions $x < 0$ and $x > L$.

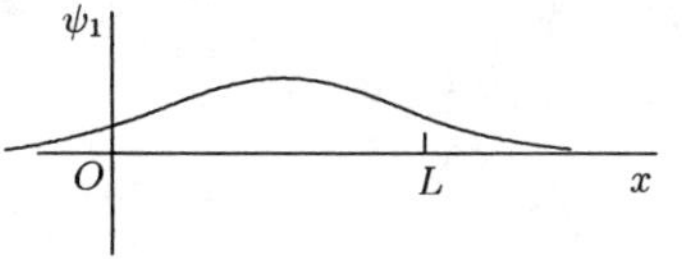

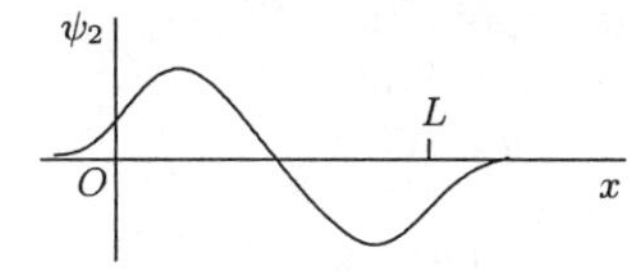

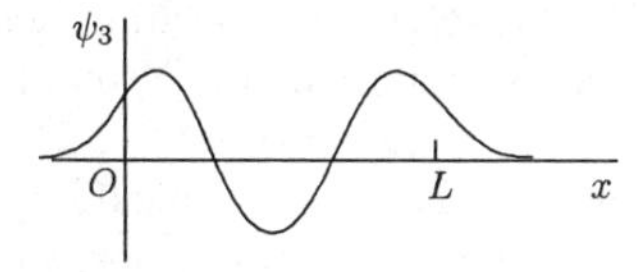

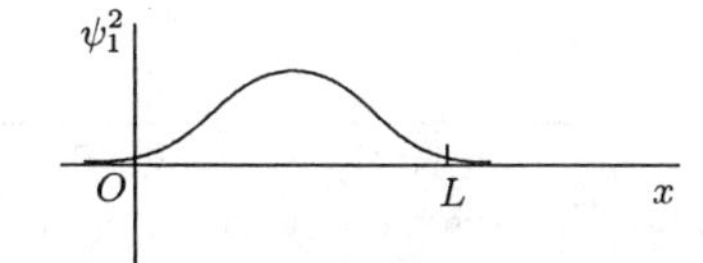

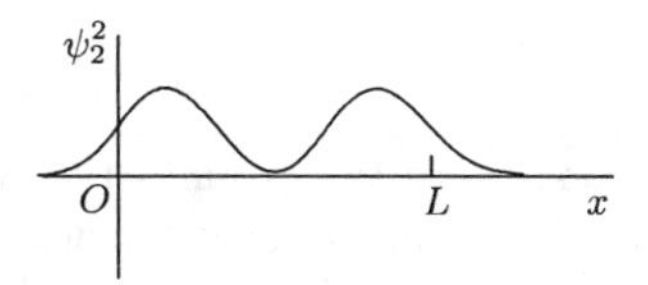

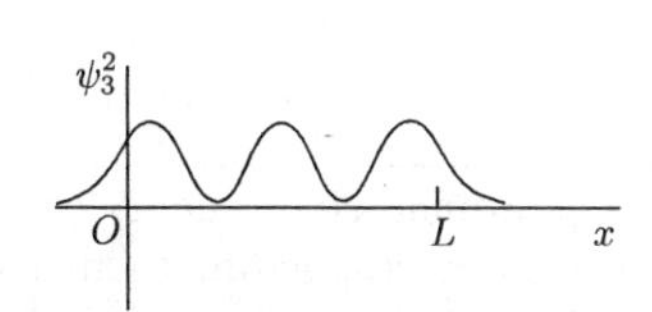

For $E < U_0$ the energies corresponding to the allowed wave functions are quantized in a similar way to the energies in an infinite well, except there are a finite number of quantized energy levels instead of an infinite number. For $E > U_0$ there is a continuum of allowed energies.

Classically, the kinetic energy of a particle is $K = E - U_0$. Because, classically, the kinetic energy must always be positive, a particle can never be found in a region where $E < U_0$. In contrast, the exponential "tail" end of the quantum mechanical results show that there is a finite probability of finding a particle in the classically forbidden regions $x < 0$ and $x > L$.

Common Pitfalls

3. TRUE or FALSE: An object can never be found in a region where its total energy is less than its potential energy.

4. What is the main difference between the energy levels of a particle in an infinite square well and in a finite square well?

35.3 The Harmonic Oscillator

In a Nutshell

An extremely important problem in quantum mechanics is the solution of the Schrödinger equation for a particle whose potential energy is that of a harmonic oscillator. This could, for example, describe small oscillations of atoms in a diatomic molecule oscillating about their equilibrium separation.

Classically, the potential energy of a particle of mass m attached to a spring of force constant k is $U(x) = \frac{1}{2}kx^2$, where x is the amount of stretch or compression of the spring from its normal unstretched position. The angular frequency of an object oscillating at the end of the spring is $\omega_0 = \sqrt{k/m}$ so that the potential energy can also be written as $U(x) = \frac{1}{2}m\omega_0^2x^2$, and is shown to the right. Classically, a particle oscillates between $x = \pm A$ with a total energy $E = \frac{1}{2}m\omega_0^2A^2$ and can have any total energy E from zero to any positive value. This is to be contrasted with the quantum mechanical results discussed next, where the total energy of a quantum-mechanical harmonic oscillator can take on only certain discrete values.

When the time-independent Schrödinger equation is solved for a harmonic oscillator potential energy function, the ground-state wave function is $\psi_0(x) = A_0\,\mathrm{e}^{-ax^2}$, where $a = m\omega_0/(2\hbar)$ and A_0 is the normalization constant. This is a Gaussian function whose plot is shown to the right.

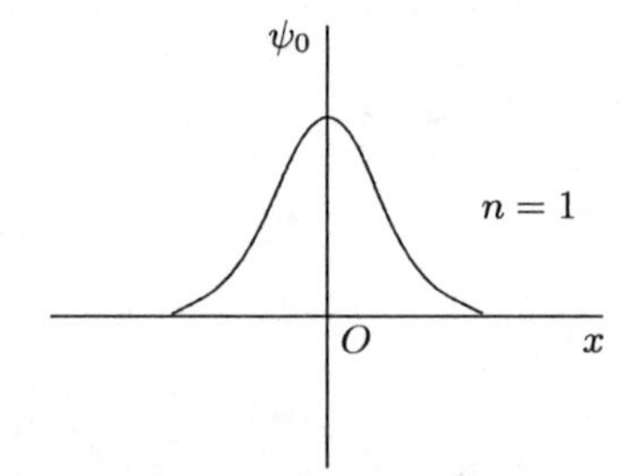

The ground-state energy corresponding to the ground-state wave function is $E_0 = \frac{1}{2}\hbar\omega_0$. The energy corresponding to higher-level wave functions is quantized and can take on only the discrete values given by $E_n = \left(n + \frac{1}{2}\right)\hbar\omega_0$, for $n = 0, 1, 2, 3, \ldots$. This expression shows that the energy levels of a quantum mechanical harmonic oscillator are evenly spaced by the amount $\hbar\omega_0$. In contrast, the energies of a classical harmonic oscillator are not quantized and can have any values.

Important Derived Results

Potential energy of a classical object attached to a spring $\quad U(x) = \dfrac{1}{2}kx^2 = \dfrac{1}{2}m\omega_0^2x^2$

Ground-state wave function for a harmonic oscillator $\quad \psi_0(x) = A_0\,\mathrm{e}^{-ax^2}$

Ground-state energy for a harmonic oscillator $\quad E_0 = \dfrac{1}{2}\hbar\omega_0$

Energy of the nth excited state for a harmonic oscillator $\quad E_n = \left(n + \dfrac{1}{2}\right)\hbar\omega_0$, for $n = 0, 1, 2, 3, \ldots$

Common Pitfalls

5. TRUE or FALSE: The ground-state wave function for a particle in a harmonic oscillator potential energy well varies sinusoidally with position.

6. Describe the energy levels associated with the harmonic oscillator potential energy well.

Try It Yourself #2

Show that the first excited wave function $\psi_1(x) = A_1 x\, e^{-ax^2}$ as shown at the right is a solution of the Schrödinger equation for a harmonic oscillator potential energy function.

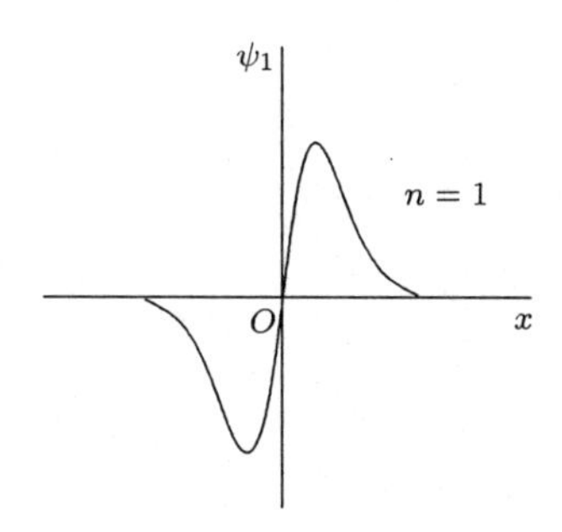

Picture: Calculate the first and second derivatives of $\psi_1(x)$ and substitute into Schrödinger's equation. If you end up with an energy that satisfies the quantization condition given in the text, then the provided wave function is a solution to Schrödinger's equation.

Solve:

Write Schrödinger's equation as a guide for the problem.	
Determine the potential energy function.	
Find the second derivative of $\psi_1(x)$.	

Substitute these values into Schrödinger's equation and solve for E.	$a = m\omega_0/(2\hbar)$ and $E_1 = \frac{3}{2}\hbar\omega_0$

Check: The energy arrived at satisfies the quantization condition given in the text, so the given wave function must be a solution to Schrödinger's equation for a harmonic oscillator.

35.4 Reflection and Transmission of Electron Waves: Barrier Penetration

In a Nutshell

A **step potential** energy function changes abruptly from zero to a finite value as shown and is represented mathematically by

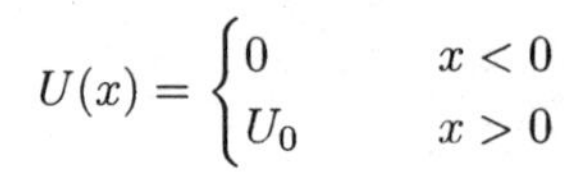

$$U(x) = \begin{cases} 0 & x < 0 \\ U_0 & x > 0 \end{cases}$$

If the energy E of the wave function incident from the left is smaller than the step potential energy U_0 most of the wave function is reflected at the original wavelength, but a small part is transmitted into the barrier and decays exponentially, as shown.

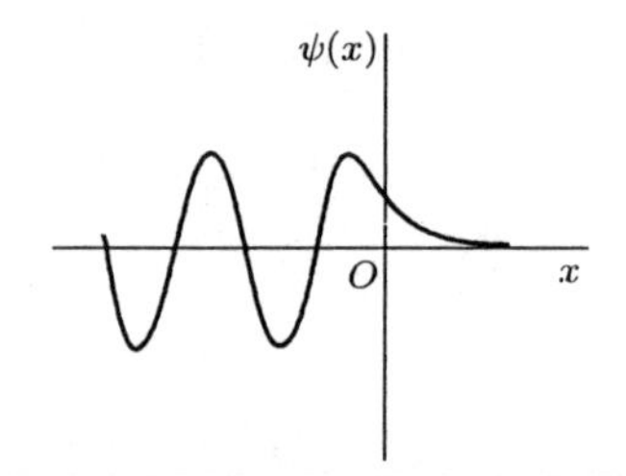

If the energy E of the incident wave is larger than the barrier potential energy U_0, part of the wave is reflected at the original wavelength and part of the wave is transmitted at a larger wavelength. The probability of reflection R, called the **reflection coefficient**, is given by $R = (k_1 - k_2)^2 / (k_1 + k_2)^2$, where k_1 and k_2 are the wave numbers for the incident and transmitted waves, respectively ($k = 2\pi/\lambda$). The probability of transmission T of the transmitted wave, the **transmission coefficient**, is related to the reflection coefficient by $T = 1 - R$.

A **barrier potential energy** is like a step potential energy except that after a certain distance a the constant potential energy U_0 changes abruptly back to zero, as shown. A barrier potential is represented mathematically by

$$U(x) = \begin{cases} 0 & x < 0 \\ U_0 & 0 < x < a \\ 0 & a < x \end{cases}$$

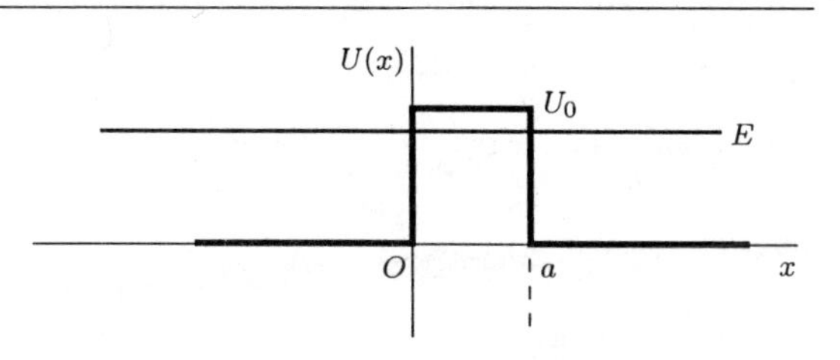

Consider a particle, represented by its wave function, incident on a barrier potential at $x = 0$ as shown. If the energy of the particle is less than the energy U_0 of the barrier energy, the wave function is partially reflected; but there also is exponential transmission of part of the wave function in the region $0 < x < a$ which is forbidden from classical energy considerations. When the exponentially transmitted wave function reaches the other end of the barrier at $x = a$, part of it is reflected again, but another part is transmitted and emerges on the other side with a reduced amplitude but with a wavelength the same as its initial wavelength.

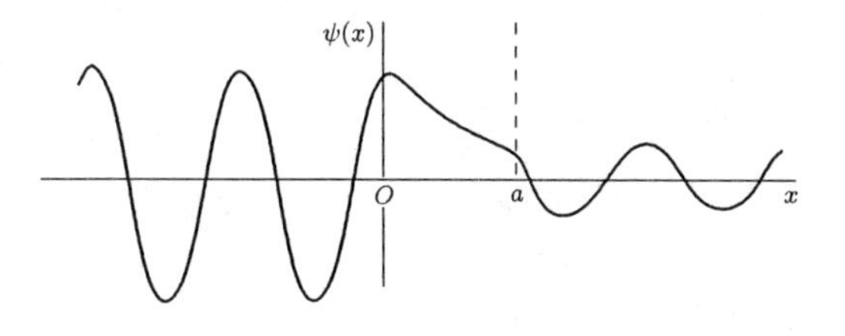

The transmission of a wave function through a barrier is referred to as **barrier penetration**. Because of the reflections and transmissions at the two sides of the barrier ($x = 0$ and $x = a$) the analysis is somewhat complicated. For the case where the barrier size a and the function $\alpha = \sqrt{2m(U_0 - E)/\hbar^2}$ that depends on the energy difference $(U_0 - E)$ are related such that $a\alpha \gg 1$, the transmission coefficient T follows an exponential expression given by $T = e^{-2\alpha a}$. The transmission coefficient gives the ratio of the probability distribution function of the transmitted wave function to that of the wave function incident on the front of the barrier: $T = P_{\text{trans}}/P_{\text{incident}}$.

Quantum-mechanical tunneling through classically forbidden regions explains phenomena such as the emission of α particles from a nucleus and the operation of a scanning tunneling microscope.

Important Derived Results

Step potential energy	$U(x) = \begin{cases} 0 & x < 0 \\ U_0 & x > 0 \end{cases}$
Reflection coefficient for a wave incident on a step potential energy	$R = \dfrac{(k_1 - k_2)^2}{(k_1 + k_2)^2}$
Transmission coefficient for a wave incident on a change in potential energy	$T = 1 - R$
Barrier potential energy	$U(x) = \begin{cases} 0 & x < 0 \\ U_0 & 0 < x < a \\ 0 & a < x \end{cases}$
Transmission coefficient through a barrier	$T = \mathrm{e}^{-2\alpha a},\ \alpha = \sqrt{\dfrac{2m(U_0 - E)}{\hbar^2}}$

Common Pitfalls

> Understand the qualitative differences among the wave functions, energy levels, and tunneling effects of the various types of potential energy functions: particle in a box; particle in a finite square well; harmonic oscillator; step potential; barrier penetration.

> Do not confuse the reflection and transmission coefficients for a particle incident on a step potential.

7. TRUE or FALSE: When an object penetrates a step potential from the left with $E < U_0$, the wavelength of the wave function on the left side of the step is smaller than the wavelength on the right side.

8. When a particle penetrates a potential barrier, on which side of the barrier is the amplitude of the wave function greater? Explain.

Try It Yourself #3

A 5.00-MeV electron strikes a 20.0-MeV potential barrier that is 10^{-13} m = 100 fm wide. Determine the probability that the electron will tunnel through the barrier.

Picture: Use the transmission coefficient for barrier penetration.

Solve:

Write the expression for the transmission coefficient to guide you through the problem.	

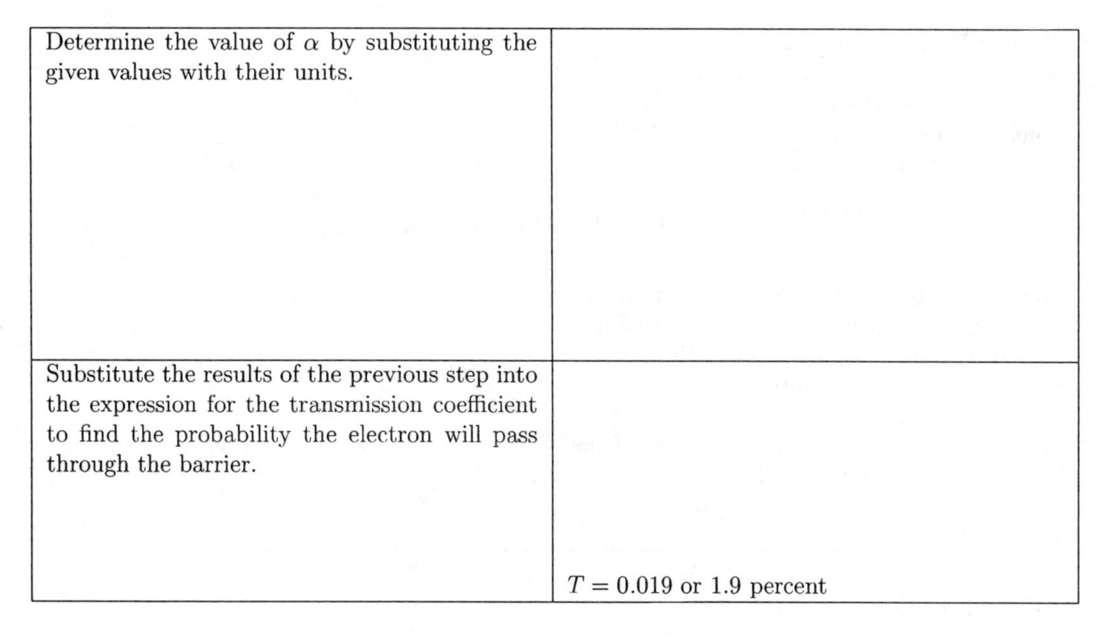

Determine the value of α by substituting the given values with their units.	
Substitute the results of the previous step into the expression for the transmission coefficient to find the probability the electron will pass through the barrier.	$T = 0.019$ or 1.9 percent

Check: The probability is less than 1, which it should be.

Taking It Further: Is it possible to alter the barrier to prevent transmission of the electron through it? Explain.

35.5 The Schrödinger Equation in Three Dimensions

In a Nutshell

In three dimensions, the time-independent Schrödinger equation in cartesian coordinates has the form

$$-\frac{\hbar^2}{2m}\left(\frac{\partial^2\psi}{\partial x^2}+\frac{\partial^2\psi}{\partial y^2}+\frac{\partial^2\psi}{\partial z^2}\right)+U\psi=E\psi$$

To see the differences between three dimensions and one dimension, imagine a particle confined to a three-dimensional cubical box of side L, in which the potential energy is

$$U(x,y,z)=\begin{cases}0 & 0<x,y,z<L\\ \infty & x,y,z \text{ outside the box}\end{cases}$$

For this three-dimensional potential, the solution to the Schrödinger equation has the form

$$\psi(x,y,z)=A\sin\frac{n_1\pi x}{L}\sin\frac{n_2\pi y}{L}\sin\frac{n_3\pi z}{L}$$

where n_1, n_2, and n_3 can each be any positive integer. The quantized energies are

$$E_{n_1,n_2,n_3}=\frac{\hbar^2\pi^2}{2mL^2}(n_1^2+n_2^2+n_3^2)$$

If more than one wave function is associated with the same energy level, that energy level is said to be **degenerate**. For example, the three combinations $(n_1 n_2 n_3) = (211); (121); (112)$ all correspond to the same energy level

$$E_{n_1 n_2 n_3} = 6\frac{\hbar^2\pi^2}{2mL^2} = 2E_{111} = 6E_1$$

where E_1 is the ground-state energy for an infinite one-dimensional square-well potential discussed previously. However, these three states describe three different wave functions, so there is a threefold degeneracy for this energy level. The degeneracy is removed if the sides of the box are made unequal, as shown for the case $L_1 < L_2 < L_3$ where $U = 0$ for $0 < x < L_1$, $0 < y < L_2$ and $0 < z < L_3$.

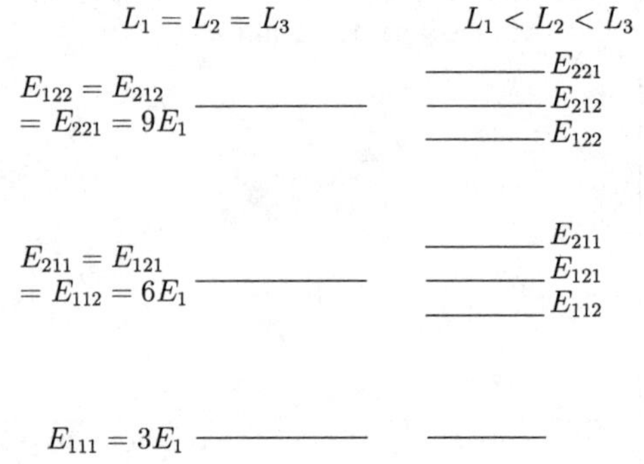

$L_1 = L_2 = L_3$	$L_1 < L_2 < L_3$
$E_{122} = E_{212} = E_{221} = 9E_1$	E_{221}, E_{212}, E_{122}
$E_{211} = E_{121} = E_{112} = 6E_1$	E_{211}, E_{121}, E_{112}
$E_{111} = 3E_1$	

Important Derived Results

Wave function for a particle in a three-dimensional box

$$-\frac{\hbar^2}{2m}\left(\frac{\partial^2\psi}{\partial x^2} + \frac{\partial^2\psi}{\partial y^2} + \frac{\partial^2\psi}{\partial z^2}\right) + U\psi = E\psi$$

Three-dimensional infinite square-well potential energy (three-dimensional box)

$$U(x,y,z) = \begin{cases} 0 & 0 < x,y,z < L \\ \infty & x,y,z \text{ outside the box} \end{cases}$$

Wave function for a particle in a three-dimensional box

$$\psi(x,y,z) = A\sin\frac{n_1\pi x}{L}\sin\frac{n_2\pi y}{L}\sin\frac{n_3\pi z}{L}$$

Energy levels for a three-dimensional box

$$E_{n_1,n_2,n_3} = \frac{\hbar^2\pi^2}{2mL^2}(n_1^2 + n_2^2 + n_3^2)$$

Common Pitfalls

9. TRUE or FALSE: Degeneracy refers to a situation where a wave function becomes infinite.

35.6 The Schrödinger Equation for Two Identical Particles

In a Nutshell

Classically, the positions of two particles can be followed through any interactions without ambiguity along well-defined distinguishable trajectories, as shown in Figures (a) and (b). In quantum mechanics, when two identical particles come together, interact, and then move apart, it is not possible after the interaction for us to know which particle is which. This is illustrated in Figure (c).

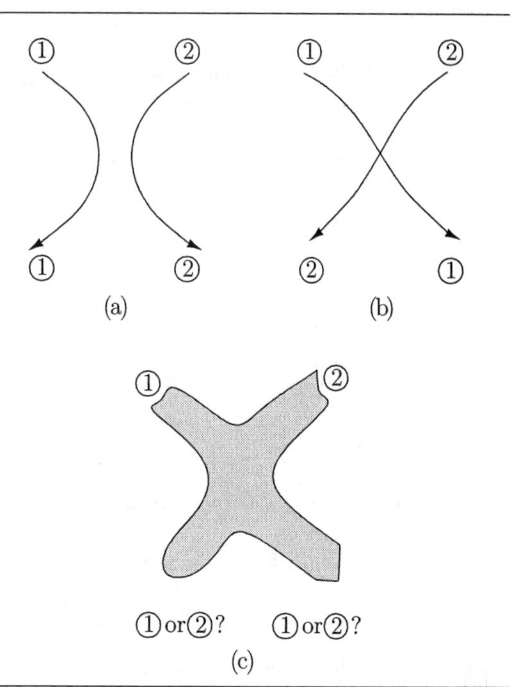

Designate by $x_1(t)$ and $x_2(t)$ the coordinates of the trajectories of two identical particles. Let $\psi(x_1, x_2)$ represent the combined wave function of two identical particles that follow one set of trajectories, and let $\psi(x_2, x_1)$ be the wave function for interchanged trajectories. The wave functions are said to be **symmetric** or **antisymmetric** upon the exchange of the two identical particles according to

$$\psi(x_1, x_2) = \psi(x_2, x_1) \qquad \text{Definition of symmetric wave function}$$
$$\psi(x_1, x_2) = -\psi(x_2, x_1) \qquad \text{Definition of antisymmetric wave function}$$

Both symmetric and antisymmetric wave functions yield the same probability densities because $\psi^2(x_1, x_2) = \psi^2(x_2, x_1)$ As an example, consider the wave function describing two identical noninteracting particles in a one-dimensional box. In Section 35-1, we saw that the wave function for one particle in a box is $\psi_n(x) = \sqrt{2/L}\sin(n\pi x/L)$. For two particles in a box, the symmetric and antisymmetric wave functions are

$$\psi_{\text{S}}(x_1, x_2) = A'\left[\psi_n(x_1)\psi_m(x_2) + \psi_n(x_2)\psi_m(x_1)\right] \qquad \text{symmetric}$$
$$\psi_{\text{A}}(x_1, x_2) = A'\left[\psi_n(x_1)\psi_m(x_2) - \psi_n(x_2)\psi_m(x_1)\right] \qquad \text{antisymmetric}$$

where A' is a constant determined by the normalization condition for the two-particle wave function.

An important difference is seen between symmetric and antisymmetric wave functions when the two quantum numbers are equal. When $n = m$ the antisymmetric wave function is identically zero for all values of x_1 and x_2, but the symmetric wave function is *not* zero for $n = m$.

A two-electron system is described by an antisymmetric wave function. If the two wave functions of the individual electrons that make up an overall antisymmetric wave function have the same quantum numbers, the overall wave function is identically zero. This is an example of the **Pauli exclusion principle** for electrons in an atom: no two electrons in an atom can have the same quantum numbers.

Besides electrons, other particles such as protons and neutrons, called **fermions**, are described by antisymmetric wave functions and obey the Pauli exclusion principle. Another class of particles, called **bosons**, which includes α particles, deuterons, photons, and mesons, are described by symmetric wave functions and do not obey the Pauli exclusion principle.

Important Derived Results

Symmetric wave functions $$\psi_S(x_1, x_2) = A'\left[\psi_n(x_1)\psi_m(x_2) + \psi_n(x_2)\psi_m(x_1)\right]$$

Antisymmetric wave functions $$\psi_A(x_1, x_2) = A'\left[\psi_n(x_1)\psi_m(x_2) - \psi_n(x_2)\psi_m(x_1)\right]$$

Common Pitfalls

- Understand the difference between symmetric and antisymmetric wave functions for two identical particles.
- Do not confuse fermions and bosons.

10. TRUE or FALSE: When a fermion is raised to a higher energy level, it becomes a boson.

11. What is the main difference between an antisymmetric and a symmetric wave function for two identical particles in the same state?

QUIZ

1. TRUE or FALSE: When two identical particles are described by a symmetric wave function, the symmetric wave function is identically zero when all the quantum numbers of each of the particles are the same.
2. TRUE or FALSE: Quantized energy levels arise because of boundary conditions on the potential energy.
3. Classically, why is a region where $E < U_0$ called a "forbidden" region?
4. What is the main difference between fermions and bosons?
5. When a particle penetrates a potential barrier, on which side of the barrier is the wavelength of the wave function greater? Explain.

6. In Try It Yourself #3, what energy of an incident electron will give it a 5 percent probability of being transmitted through the barrier?

7. A particle is confined to a two-dimensional box defined by the following boundary conditions: $U(x, y) = 0$ for $-L/2 \leq x \leq L/2$ and $-3L/2 \leq y \leq 3L/2$, and $U(x, y) = \infty$ outside these ranges. Determine the energies of the three lowest energy states and their degeneracies.

Chapter 36

Atoms

36.1 The Atom

In a Nutshell

Each element is characterized by a neutral atom that contains a number of protons Z, an equal number of electrons, and a number of neutrons N. The protons and neutrons are in the nucleus, and the electrons orbit the nucleus. The chemical and physical properties of an element are determined by the number and arrangement of the electrons in the atom.

By the end of the nineteenth century, the emission spectra of many gaseous elements were known quite accurately. In 1885, Balmer determined that the wavelengths of the lines in the visible spectrum of hydrogen were given by

$$\lambda = (364.6 \text{ nm})\frac{m^2}{m^2 - 4} \quad \text{for } m = 3, 4, 5, \ldots$$

Rydberg and Ritz extended Balmer's expression, arriving at the **Rydberg–Ritz formula**, which holds for all alkali metals:

$$\frac{1}{\lambda} = R\left(\frac{1}{n_2^2} - \frac{1}{n_1^2}\right)$$

Here R is the Rydberg constant, which actually varies slightly, but regularly, from element to element. For hydrogen, $R_{\mathrm{H}} = 1.09776 \times 10^7 \text{ m}^{-1}$.

Initial models to explain the above results included Thomson's "plum pudding" model, in which positive charges were uniformly distributed throughout the entire volume of an atom, which was known to have a diameter of about 0.1 nm. Geiger, Marsden, and Rutherford ruled out this model with experiments that demonstrated that most of the mass, and the positive charge of the atom, were concentrated in a region with a diameter of only about 10^{-6} nm = 1 fm.

Physical Quantities and Their Units

Rydberg constant for hydrogen	$R_{\mathrm{H}} = 1.09776 \times 10^7 \text{ m}^{-1}$

Important Derived Results

Rydberg–Ritz formula	$\frac{1}{\lambda} = R\left(\frac{1}{n_2^2} - \frac{1}{n_1^2}\right)$

36.2 The Bohr Model of the Hydrogen Atom

In a Nutshell

In 1913, Niels Bohr put forth a model of the hydrogen atom that was in extraordinary agreement with the Rydberg–Ritz formula. Bohr's model is based on a planetary picture of a hydrogen-like atom consisting of a light electron moving in a circular orbit around a heavy nucleus. The attractive Coulomb force $F_C = kZe^2/r^2$ between the orbiting electron and Z protons in the nucleus produces the necessary centripetal force for the electron to move in the circular orbit. The total energy of this circular orbit is given by $E = -kZe^2/(2r)$. Bohr's model is built on three postulates.

Bohr's First Postulate: Nonradiating Orbits (Stationary States). According to classical electromagnetic theory, an electron that experiences centripetal acceleration in a circular orbit should radiate electromagnetic energy (light). This continuous loss of energy should cause the electron to spiral in to the nucleus. But there is no such loss of energy observed, and the electron does not spiral inward. Bohr's "solution" to this dilemma was to postulate that an electron can exist only in discrete stable orbits, called stationary states, and in these states there is no radiation.

Bohr's Second Postulate: Photon Frequency from Energy Conservation (Radiative Transitions). Bohr built his second postulate on the quantum ideas of Planck and Einstein, postulating that photons are emitted when the electron in a hydrogen-like atom suddenly changes from one stable orbit to another, in what he called a radiative transition. The energy of an emitted photon $E = hf$ equals the difference between the energy of an electron in the initial orbit E_i and its energy in the final orbit E_f, giving the frequency of an emitted photon as $f = (E_\mathrm{i} - E_\mathrm{f})/h$.

Bohr's Third Postulate: Quantized Angular Momentum. The energy of an electron in one of its stable orbits is quantized in Bohr's third postulate: the angular momentum of an electron in an orbit can take on only certain discrete values given by $mvr = n\hbar$, where $n = 1, 2, 3, \ldots$.

Application of Bohr's three postulates to a hydrogen-like atom gives the radii of the allowed orbits of the electron as $r_n = n^2\hbar^2/(mkZe^2) = n^2a_0/Z$, where again $n = 1, 2, 3, \ldots$, k is the Coulomb constant, and **the first Bohr radius** (when $n = 1$) is $a_0 = \hbar^2/(mke^2) \approx 0.0529$ nm.

The quantized energy E_n of an electron in one of the quantized orbits is $E_n = -mk^2e^4Z^2/(2\hbar^2n^2) = -Z^2E_0/n^2$, where $n = 1, 2, 3, \ldots$. For hydrogen ($Z = 1$) the lowest energy, $-E_0$, called the **ground-state energy**, corresponds to $n = 1$ and is given by $E_0 = -mk^2e^4/(2\hbar^2) = -ke^2/(2a_0) \approx -13.6$ eV. The magnitude of the ground-state energy for hydrogen-like atoms is given by $E_\mathrm{g} = Z^2E_0$.

The Rydberg–Ritz formula can be derived from Bohr's second postulate. Comparison of the expression derived from Bohr's second postulate with the Rydberg–Ritz formula shows that the experimentally determined Rydberg constant can be expressed in terms of fundamental constants of nature: $R = mk^2e^4/(4\pi\hbar^3c)$. When R is evaluated by replacing these constants with their numerical values, very close agreement with the experimental spectroscopic value of R is obtained; this is a major success of the Bohr model.

The specification of n_2 in the Rydberg–Ritz formula systematically defines series of spectral lines that lie in the visible and nonvisible regions according to the following scheme (each series is named for the person who first observed it experimentally for hydrogen):

$n_2 = 1$, $n_1 = 2, 3, 4, \ldots$ Lyman series (ultraviolet region)
$n_2 = 2$, $n_1 = 3, 4, 5 \ldots$ Balmer series (optical region)
$n_2 = 3$, $n_1 = 4, 5, 6 \ldots$ Paschen series (near-infrared region)
$n_2 = 4$, $n_1 = 5, 6, 7, \ldots$ Brackett series (infrared region)

A standard way to represent the energies is with an energy-level diagram, such as the one shown for hydrogen, in which the transitions between energy levels E_n are shown as arrows. When an electron occupies an energy level above its ground state ($n = 1$), it is said to be in an **excited state**.

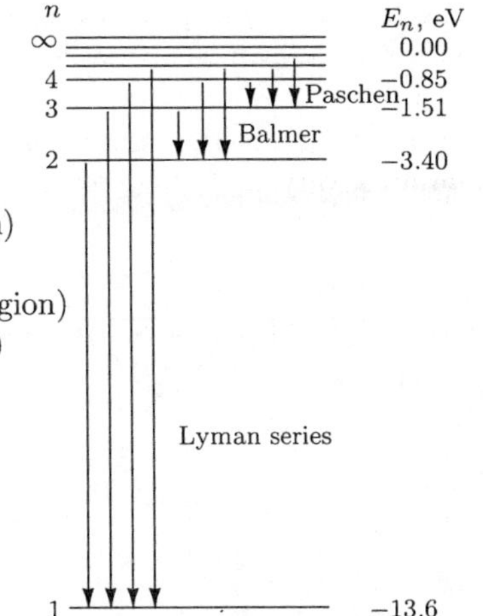

The **ionization energy** is the energy necessary to remove an electron from an atom. For a hydrogen atom, this is the energy required to remove an electron from the ground state ($n = 1$) to zero energy ($n = \infty$), which is 13.6 eV.

Physical Quantities and Their Units

$E_n = (n + 1/2)\hbar\omega$

General Rydberg constant $$R = \frac{mk^2e^4}{4\pi\hbar^3 c}$$

First Bohr radius $$a_0 = \frac{\hbar^2}{mke^2} \approx 0.0529 \text{ nm}$$

Ground-state energy for hydrogen $$E_0 = -\frac{mk^2e^4}{2\hbar^2} = -\frac{ke^2}{2a_0} \approx -13.6 \text{ eV}$$

Important Derived Results

Frequency of a photon emitted in the Bohr model $$f = \frac{E_i - E_f}{h}$$

Radii of allowed orbits $$r_n = \frac{n^2\hbar^2}{mkZe^2} = n^2\frac{a_0}{Z}$$

Quantized energy levels for hydrogen-like atoms $$E_n = -\frac{mk^2e^4Z^2}{2\hbar^2 n^2} = -Z^2\frac{E_0}{n^2}$$

Bohr expression for $1/\lambda$ $$\frac{1}{\lambda} = Z^2\frac{mk^2e^4}{4\pi\hbar^3 c}\left(\frac{1}{n_2^2} - \frac{1}{n_1^2}\right)$$

Common Pitfalls

> A main result of the Bohr model of hydrogen-like atoms is that the energy of the atom is quantized. Make sure you understand that the energy $E = hf$ of an emitted photon equals the difference of the energies of the initial energy level E_i and final energy level E_f between which an electron makes a transition: $hf = E_i - E_f$.

1. TRUE or FALSE: The absolute value of the ionization energy of a hydrogen-like atom is numerically equal to the absolute value of the energy of the energy level.

2. How do the energies of an electron in the Bohr model vary with n?

$E_n = -Z^2\left(\frac{E_0}{n^2}\right) \rightarrow$ w/ $1/n^2$

Try It Yourself #1

The wavelengths of visible light range from about 400 to 700 nm. Determine the wavelengths in the Balmer series for hydrogen that lie in the visible region.

Picture: Determine the values of n_2 and Z that pertain to this problem. Use them in the Rydberg–Ritz formula, varying n_1 to determine the wavelengths in the visible range.

Solve:

Determine Z and n_2 for the Balmer series in hydrogen.	$Z = 1$, $n_2 = 2$
Write the Rydberg–Ritz formula and substitute the above values to reduce the formula to a simple form in terms of only n_1.	$\frac{1}{\lambda} = R\left(\frac{1}{4} - \frac{1}{n_1^2}\right)$ ↳ 1.0977 E 7
Determine the longest wavelength in the Balmer series by setting $n_1 = 3$. This wavelength is in the visible spectrum.	$n_1 = 3$ $\frac{1}{\lambda} = (1.0977E7)\left(\frac{1}{4} - \frac{1}{9}\right)$ longest
Determine the shortest wavelength in the Balmer series by setting $n_1 = \infty$. This wavelength is outside the visible spectrum, so we must determine the largest value for n_1 in a different fashion.	$\frac{1}{\lambda} = (1.0977E7)\left(\frac{1}{4} - \frac{1}{\infty}\right)$ shortest
Determine the largest value of n_1 that produces an emission line in the visible by setting $\lambda = 400$ nm and solving for n_1. Since n_1 must be an integer, you need to round down.	$\frac{1}{400\text{nm}} = (1.0977E7)\left(\frac{1}{4} - \frac{1}{n_1^2}\right)$

Determine all the visible emission wavelengths, using values of n_1 from 3 to the value you found in the previous step.	$\lambda_3 = 656.5$ nm, $\lambda_4 = 486.3$ nm, $\lambda_5 = 434.2$ nm, $\lambda_6 = 410.3$ nm

Check: The wavelengths get shorter as the energy, represented by the value of n_1, increases, as expected

Taking It Further: If you let $n_1 = 7$, what wavelength do you get for the emission?

Try It Yourself #2

Electrons in singly ionized helium are excited into the $n = 3$ energy level. Using the Bohr model, determine the wavelengths of photons that are emitted when the electrons return to the ground state.

Picture: These excited electrons can either return directly to the ground state or first decay to the $n = 2$ state en route to the ground state.

Solve:

Determine Z, n_1 and n_2 for all possible transitions for this system.	$n = 3 \rightarrow n = 1$ $\frac{1}{\lambda} = R_H\left(\frac{1}{1} - \frac{1}{9}\right)$ $(1.0977E7)$
Write the Rydberg–Ritz formula and substitute the appropriate values to determine the emission wavelength for a direct transition from the $n = 3$ state to the $n = 1$ state.	$\lambda_{31} = 25.6$nm $\lambda_{31} = 25.6$ nm

Write the Rydberg–Ritz formula and substitute the appropriate values to determine the emission wavelength for a direct transition from the $n = 3$ state to the $n = 2$ state.	$\frac{1}{\lambda} = R_H \left(\frac{1}{9} - \frac{1}{4} \right)$ $\lambda_{32} = 164$ nm
Write the Rydberg–Ritz formula and substitute the appropriate values to determine the emission wavelength for a direct transition from the $n = 2$ state to the $n = 1$ state.	$\lambda_{21} = 30.4$ nm

Check: The energy of the $n = 3$ to $n = 1$ transition should be larger than the other two energies, so the wavelength of that transition should be shorter than the wavelengths of the other two transitions, which we found.

36.3 Quantum Theory of Atoms

In a Nutshell

Three-dimensional problems involving electrons orbiting central nuclei are most easily treated in terms of spherical coordinates (r, θ, ϕ). By applying the transformations $z = r\cos\theta$, $x = r\sin\theta\cos\phi$ and $y = r\sin\theta\sin\phi$, we can arrive at the time-independent Schrödinger equation:

$$-\frac{\hbar^2}{2m}\frac{1}{r^2}\frac{\partial}{\partial r}\left(r^2\frac{\partial\psi}{\partial r}\right) - \frac{\hbar^2}{2mr^2}\left[\frac{1}{\sin\theta}\frac{\partial}{\partial\theta}\left(\sin\theta\frac{\partial\psi}{\partial\theta}\right) + \frac{1}{\sin^2\theta}\frac{\partial^2\psi}{\partial\phi^2}\right] + U(r)\psi = E\psi$$

The solution to this equation requires writing $\psi(r, \theta, \phi) = R(r)f(\theta)g(\phi)$: that is, the wave function is **separable**—it can be written as a product of three functions (a radial function $R(r)$, a polar function $f(\theta)$, and an azimuthal function $g(\phi)$, each of which depends only on one variable.

When the Schrödinger equation in three dimensions is solved using spherical coordinates, and when electron spin is taken into consideration, it is found that there are four quantum numbers that are needed to describe the resultant wave function ψ for electrons in an atom: the principal quantum number n, the orbital quantum number ℓ, the magnetic quantum number m_ℓ, and the intrinsic spin quantum number m_s. Specifying these quantum numbers defines an electron's state.

The **principal quantum number** n, which takes on the integral values $n = 1, 2, 3, \ldots$, is roughly analogous to the quantum number n, discussed in Chapter 35, that determines wave functions and energy levels for a particle in a one-dimensional box. The principal quantum number n specifies the radial part of the wave function, which determines the probability of finding an electron at a radial distance r from the center of an atom. Increasing values of n correspond to increasing values of energy of the quantized energy levels. Electron-shell letters are related to the values of n according to the following scheme.

value of n:	1	2	3	4	...
shell letter:	K	L	M	N	...

The **orbital quantum number** is designated by ℓ. In a classical picture, ℓ is related to the angular momentum of an electron as it orbits the nucleus of an atom. It takes on integral values with a range related to the principal quantum number n according to $\ell = 0, 1, 2, \ldots, n-1$. As an electron orbits a nucleus, its orbital angular momentum L is related to its orbital quantum number ℓ by $L = \sqrt{\ell(\ell+1)}\hbar$. For the ground state $n = 1$, $\ell = 0$ and the orbital angular momentum L is zero, whereas for $n \geq 2$ there are several values of L. An orbital letter code is used to indicate the value of ℓ for an electron.

value of ℓ:	0	1	2	3	4	...
letter symbol:	s	p	d	f	g	...

If an atom is placed in an external magnetic field whose direction is taken as the z direction, we can speak of the z component L_z of an electron's orbital angular momentum $\vec{L}$ along this direction. Quantum-mechanical analysis shows that this component L_z can take on only discrete quantized values given by $L_z = m_\ell \hbar$, where m_ℓ is an integer called the **magnetic quantum number** that can have only the values $m_\ell = (-\ell), (-\ell+1), (-\ell+2), \ldots, -1, 0, 1, \ldots, (\ell-1), (\ell)$. For a given value of ℓ the magnitude of the orbital angular momentum is $L = \sqrt{\ell(\ell+1)}\hbar$, whereas the maximum value of the z component is $\ell\hbar$, which is less than L. This means that the angular momentum vector $\vec{L}$ never points exactly along the z direction. The angle θ that $\vec{L}$ makes with the z direction takes on only discrete values. The relationships among ℓ, L, m_ℓ and L_z apply not only to an individual electron in an atom, but also to the net values for the atom as a whole.

In addition to its orbital angular momentum, an electron possesses an intrinsic angular momentum $\vec{S}$ called its **spin**. The magnitude of the spin angular momentum S is quantized according to $S = \sqrt{s(s+1)}\hbar$, where $s = \frac{1}{2}$ is the electron's **intrinsic spin quantum number**. Just as L_z is quantized with the values $L_z = m_\ell \hbar$, the z component of an electron's spin is quantized according to $S_z = m_s \hbar$, where the two values of S_z are usually referred to as "spin up" for the electron spin quantum number $m_s = \frac{1}{2}$ and "spin down" for $m_s = -\frac{1}{2}$. However, the spin vector $\vec{S}$ never points exactly in the $+z$ or $-z$ direction. The classical picture of an electron is a ball of charge spinning with an intrinsic angular momentum $\vec{S}$ as it orbits the nucleus with an orbital angular momentum $\vec{L}$ much like a spinning planet orbiting the sun.

Important Derived Results

Quantum numbers for spherical coordinates

- n principal quantum number
- ℓ orbital quantum number
- m_ℓ magnetic quantum number
- m_s intrinsic spin quantum number

Orbital quantum number range $\quad \ell = 0, 1, 2, \ldots, n-1$

Orbital angular momentum $\quad L = \sqrt{\ell(\ell+1)}\hbar$

Magnetic quantum number range

$$m_\ell = (-\ell), (-\ell+1), (\ell+2), \ldots, -1, 0, 1, \ldots, (\ell+1), (\ell)$$

Common Pitfalls

> Do not confuse angular momentum, which is a physical quantity, with the quantum number that is associated with that angular momentum, which is an integer or half-integer number. For example, the value of the magnitude of the orbital angular momentum $\vec{L}$ of an atom is related to the associated quantum number ℓ by $L = \sqrt{\ell(\ell+1)}\hbar$.

> Do not confuse the orbital angular momentum with the orbital magnetic moment. The orbital angular momentum is associated with the mass of an electron orbiting a nucleus, much like the angular momentum of a massive planet as it orbits the sun. The orbital magnetic moment

is associated with the magnetic field produced by the charge of an electron that is orbiting the nucleus.

> For reasons still not fully understood by me, mathematicians and physicists switch the roles of the angles θ and ϕ. For physicists, θ is the angle with respect to the z axis, but mathematicians denote this angle as ϕ.

3. TRUE or FALSE: The angular momentum L is related to the orbital quantum number ℓ by $L = \ell\hbar$.

Try It Yourself #3

How many electrons can fit into the L shell of an atom?

Picture: The value of n can be determined from the letter code. This determines the possible values of ℓ, which in turn determine the values for m_ℓ.

Solve:

Determine the value of n.	$L \to n = 2$
Determine the number of allowed values for the ℓ quantum number.	$\ell = [0, n-1] \Rightarrow \ell_{max} = 1$
Determine the allowed values for m_ℓ for each possible value of ℓ.	$m_{\ell} \to_{max} \ell \to 1 \,\&\, 0$
For each value of m_ℓ there are two allowed spin states.	$m_\ell \to 0 - 2$ $m_\ell \to -1, 0, 1 \to 2 + 2 + 2$ } (8) 8 total states

Check: There can be two electrons in the s state ($\ell = 0$), and six electrons in the p state ($\ell = 1$).

36.4 Quantum Theory of the Hydrogen Atom

In a Nutshell

A hydrogen atom consists of a heavy nucleus composed of a proton around which a light electron orbits. The potential energy that goes into the Schrödinger equation is the Coulomb potential energy of the electron-nucleus system in the hydrogen-like atom: $U = -kZe^2/r$. The atomic number Z, which is 1 for hydrogen, is included in our expressions so the results are applicable to other hydrogen-like, one-electron atoms such as singly ionized He^+, doubly ionized Li^{2+}, and so forth. The Schrödinger equation for the hydrogen atom can be solved exactly, and the solution is denoted as $\psi_{n\ell m}$, where n, ℓ and m represent the quantum numbers of the particular state.

Upon solving the Schrödinger equation for the hydrogen atom Coulomb potential, it is found that the **allowed energy levels** are quantized according to $E_n = -mk^2e^4Z^2/(2\hbar^2n^2) = -Z^2E_0/n^2$, where $n = 1, 2, \ldots$ and $E_0 = mk^2e^4/(2\hbar^2) = ke^2/(2a_0) \approx 13.6\,\text{eV}$. Amazingly, these energy levels that arise naturally from solution of the Schrödinger equation are exactly the same quantized energies that result from application of the ad hoc Bohr theory.

The ground state of a hydrogen atom corresponds to $(n, \ell, m_\ell) = (1, 0, 0)$. The normalized wave function has the form $\psi_{100} = (1/\sqrt{\pi})(Z/a_0)^{3/2}\,e^{-Zr/a_0}$, where, remarkably, the solution of the Schrödinger equation gives $a_0 = \hbar^2/(mke^2) = 0.0529$ nm, which is equal to the first Bohr radius obtained from Bohr's theory.

Note that ψ_{100} depends only on the radial coordinate r, with no dependence on the angles θ or ϕ. Also note that since $\ell = 0$, the correct orbital angular momentum is zero, in contrast to the Bohr model, which incorrectly gives the ground-state angular momentum as $\hbar$.

Because the ground-state wave function has no angular dependence, we do not have to consider angles in determining the probability of finding an electron in the three-dimensional space around the nucleus of the hydrogen-like atom. Thus we are interested in the radial probability density $P(r)$ for finding an electron between radii r and $r + dr$. To find this, take the volume element of a spherical shell of thickness dr, which is $dV = 4\pi r^2\,dr$, and multiply it by the (volume) probability density ψ_{100}^2 to get the probability of finding an electron in the spherical shell: $P(r)\,dr = 4\pi r^2\psi_{100}^2\,dr$, resulting in the radial probability density of $P(r) = 4\pi r^2\psi_{100}^2 = 4(Z/a_0)^3r^2\,e^{-2Zr/a_0}$, which is shown here.

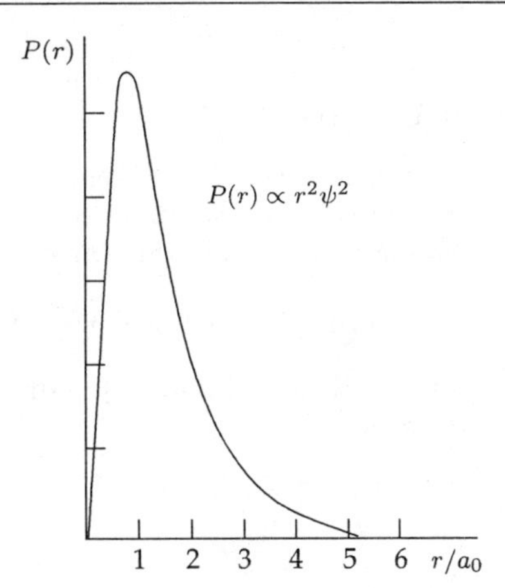

For $Z = 1$ the maximum value of $P(r)$ occurs exactly at the first Bohr radius $r = a_0$, showing that the Schrödinger equation predicts that this is the most likely place to find the electron in the ground state of a hydrogen atom. Note, though, the difference between the Bohr model and the Schrödinger theory for the ground state of hydrogen. The Bohr-model electron stays exactly in a well-defined circular orbit of radius a_0. In the wave-mechanical picture, the most likely place to find a quantum-theory electron is at the radius a_0, but there are finite probabilities for finding it at other radial distances from the nucleus.

The first excited states of hydrogen correspond to $n = 2$, and the wave functions have the form

$$\psi_{200} = C_{200}\left(2 - \frac{Zr}{a_0}\right) e^{-Zr/2a_0}$$

$$\psi_{210} = C_{210}\frac{Zr}{a_0} e^{-Zr/2a_0} \cos\theta$$

$$\psi_{21\pm1} = C_{21\pm1}\frac{Zr}{a_0} e^{-Zr/2a_0} \sin\theta\, e^{\pm i\phi}$$

where $C_{n\ell m}$ is a constant to be determined from the normalization condition.

Important Derived Results

Ground-state hydrogen wave function
$$\psi_{100} = \frac{1}{\sqrt{\pi}}\left(\frac{Z}{a_0}\right)^{3/2} e^{-Zr/a_0}$$

Ground-state radial probability density for hydrogen
$$P(r) = 4\pi r^2 \psi_{100}^2 = 4\left(\frac{Z}{a_0}\right)^3 r^2\, e^{-2Zr/a_0}$$

First excited state hydrogen atom wave functions
$$\psi_{200} = C_{200}\left(2 - \frac{Zr}{a_0}\right) e^{-Zr/2a_0}$$
$$\psi_{210} = C_{210}\frac{Zr}{a_0} e^{-Zr/2a_0} \cos\theta$$
$$\psi_{21\pm1} = C_{21\pm1}\frac{Zr}{a_0} e^{-Zr/2a_0} \sin\theta\, e^{\pm i\phi}$$

Common Pitfalls

> For the ground state of hydrogen, the probability density ψ_{100}^2 is spherically symmetric, with a maximum value at the center of the atom. However, the radial probability density $P(r) = 4\pi r^2 \psi_{100}^2$ has a maximum value at $r = a_0$. Understand the difference between these functions.

4. TRUE or FALSE: The hydrogen wave functions with $n = 2$ are all spherically symmetric.

5. Why doesn't the radial probability distribution function $P(r)$ for the ground state of hydrogen simply equal $\psi^2(r)$?

$\psi^2(r)\,dV \to$ is volumetric

Try It Yourself #4

For $n = 2$, $\ell = 0$, the hydrogen wave function is spherically symmetric and has the form $\psi_{200} = C_{200}(2 - r/a_0)\, e^{-r/2a_0}$. Find the location(s) of r/a_0 where the radial probability density function is a maximum.

Picture: Find the probability density in terms of the wave function. Maxima and minima are located where the derivative of the probability density function is zero.

Solve:

Determine the probability density function.	$P(r) = Cr^2e^{-2Zr/a_0}$
Find the derivative of the probability density with respect to r.	
Set the derivative equal to zero and solve for r.	
Using the second derivative test, or plotting $P(r)$, determine which values of r/a_0 correspond to maximum values of $P(r)$.	$r/a_0 = 0.764, 5.24$

Check: These values correspond to radii less than $10a_0$, which seems reasonable.

Taking It Further: By plotting them both, compare the probability density for the $n = 2$, $\ell = 1$ state to that found here.

36.5 The Spin–Orbit Effect and Fine Structure

In a Nutshell

The **total angular momentum** $\vec{J}$ of an atom is equal to the vector sum of the orbital angular momentum $\vec{L}$ and the spin angular momentum $\vec{S}$: $\vec{J} = \vec{L} + \vec{S}$. It follows from the vector addition of the orbital and spin angular momenta that the magnitude of $\vec{J}$ is quantized according to the rule $|\vec{J}| = \sqrt{j(j+1)}\hbar$, where $j = \frac{1}{2}$ for $\ell = 0$, and $j = \ell \pm \frac{1}{2}$ for $\ell > 0$.

The vector addition $\vec{J} = \vec{L} + \vec{S}$ for the case $\ell = 1$ is shown here. For this case j has the two values $j = \ell - \frac{1}{2} = \frac{1}{2}$ and $j = \ell + \frac{1}{2} = \frac{3}{2}$. The lengths of the various angular momentum vectors are $L = \sqrt{2}\hbar$, $S = (\sqrt{3}/2)\hbar$, and $J = (\sqrt{15}/2)\hbar$ or $J = (\sqrt{3}/2)\hbar$.

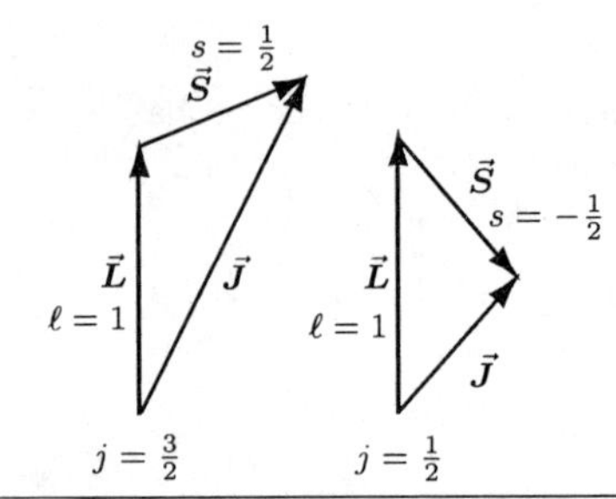

The notation for specifying the angular momentum quantum numbers, j, ℓ, and s for an atom is to use capital letters S, P, D, and F to designate ℓ for an atomic state, and lower-case letters s, p, d, and f to designate ℓ for the states of individual electrons in the atom. The letter value for ℓ is preceded by $2s + 1$ as a superscript and followed by j as a subscript. Thus, an atom in a state $^2P_{3/2}$ has quantum numbers $\ell = 1$, $s = \frac{1}{2}$, and $j = \frac{3}{2}$. Sometimes the pre-superscript is omitted, and the letter symbol is multiplied by the value of n of the shell being considered, to get the notation nl_j. For example, the states $2P_{3/2}$ and $2P_{1/2}$ correspond to a doublet energy level in which $n = 2$, $\ell = 1$, and $j = \frac{3}{2}$ and $j = \frac{1}{2}$, with $s = \frac{1}{2}$ implied.

An electron orbiting a nucleus has an orbital angular momentum $\vec{L}$ and a spin angular momentum $\vec{S}$. Associated with the spin angular momentum is a spin magnetic moment $\vec{\mu}_s$. From the point of view of the electron, the positively charged nucleus orbits around the electron, thereby producing a magnetic field $\vec{B}$ at the location of the electron. The spin magnetic moment $\vec{\mu}_s$ placed in the orbital magnetic field $\vec{B}$ gives rise to a potential energy $U = -\vec{\mu}_s \cdot \vec{B}$. Thus the orbital magnetic field interacts with the spin magnetic moment of the electron, giving the electron an additional potential energy that is added or subtracted from its energy without the spin–orbit effect.

The net result of this **spin–orbit effect** is that an energy level for which $\ell \neq 0$ is split into sublevels. For atoms such as hydrogen or sodium, which have one outermost electron, there are two sublevels corresponding to the two values $s = +\frac{1}{2}$ and $s = -\frac{1}{2}$, with the $j = \ell + \frac{1}{2}$ level being slightly higher than the $j = -\frac{1}{2}$ level, as shown. Because spin–orbit coupling produces closely spaced energy levels for each value of $\ell \neq 0$, transitions from these levels to lower levels result in the emission of photons with slightly different wavelengths, as shown. This is called **fine structure**, which is easily observed with spectrometers of moderate resolution.

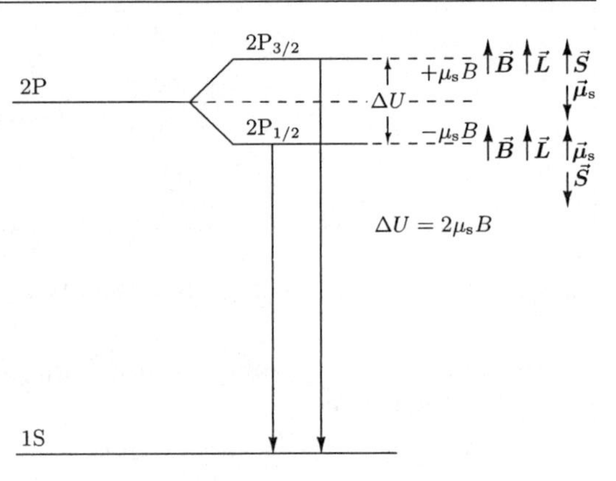

Common Pitfalls

> It can be easy to confuse the orbital angular momentum with the orbital magnetic moment. The orbital angular momentum is associated with the mass of an electron orbiting a nucleus, much like the angular momentum of a massive planet as it orbits the sun. The orbital magnetic moment is associated with the magnetic field produced by the charge of an electron that is orbiting the nucleus.

> Recognize the distinction between the spin angular momentum and the spin magnetic moment. The spin angular momentum is associated with the mass of an electron rotating about an axis, much like the angular momentum of the earth rotating about its axis. The spin magnetic moment is associated with the magnetic field produced by the charge of the electron as it spins.

6. TRUE or FALSE: In a doublet, $j = \ell - \frac{1}{2}$ energy levels are lower than $j = \ell + \frac{1}{2}$ energy levels.

7. Most spectral lines of atoms occur in groups of two or more closely spaced lines. What causes these groupings?

Spin orbit coupling. $j = \ell \pm 1/2$

Try It Yourself #5

In a particular atom the two outermost electrons in p states combine to give $\ell = 1$ and $s = 1$. Give the spectroscopic notation of the possible states of the atom.

Picture: Spectroscopic notation is given by $^{2s+1}\ell_j$.

Solve:

Determine the letter code for $\ell = 1$.	$\ell = 0 \to s$, $\ell = 1 \to p$
Determine the allowed values of j.	$j = \ell \pm 1/2 \to j = 1/2, 3/2$
Convert all allowed states to spectroscopic notation.	$^3P_{3/2}$, $^3P_{1/2}$ $^3P_2$, $^3P_1$, $^3P_0$

Try It Yourself #6

What are the quantum numbers for an atom in a $^1D_2$ state?

Picture: Spectroscopic notation corresponds to the quantum numbers according to $^{2s+1}\ell_j$.

Solve:

Determine the value of s from the preceding superscript value.	1 = 2s+1 ⇒ s=0 $s = 0$
Determine the value of $\ell = 1$ from the letter code.	D → ℓ=2 $\ell = 2$
Determine the value of j from the subscript.	j =2 $j = 2$

36.6 The Periodic Table

In a Nutshell

The properties and ordering of elements in the periodic table derive largely from application of the Pauli exclusion principle, which states that no two electrons in an atom can have the exact same set of all four quantum numbers (n, ℓ, m_ℓ, m_s). As a consequence, as Z increases from one ground-state atom to the next, the trend is for each added electron to be found in a higher energy level.

As mentioned earlier, the quantum number n designates a shell of electrons in an atom. The pair (n, ℓ) specifies a subshell. The Pauli exclusion principle determines how many electrons can occupy a given subshell. There are $2\ell + 1$ values of m_ℓ for each value of ℓ. Because there are two values of m_s ($+\frac{1}{2}$ and $-\frac{1}{2}$) for each value of m_ℓ the maximum number of electrons that can occupy a subshell is $2(2\ell + 1)$. This allows us to construct a table for specific values of ℓ:

ℓ value	0	1	2	3
Subshell letter	s	p	d	f
Maximum number of electrons	2	6	10	14

The subshells of most (but not all) atoms fill from bottom to top in the order shown, which also gives a rough indication of the relative values of the energy of each level. When a shell completely fills with electrons, there is a relatively large energy gap to the next subshell. These filled shells correspond to the noble or inert gases at $Z = 2$, 10, 18, 36, 54, and 86, which are difficult to ionize and chemically inactive. Other chemical properties of the elements also depend on the various ways their subshells are filled.

Shell	Level	Maximum number of electrons	Total number of electrons
P	6p	6	86
	5d	10	
	4f	14	
	6s	2	
O	5p	6	
	4d	10	54
	5s	2	
N	4p	6	
	3d	10	36
	4s	2	
M	3p	6	18
	3s	2	
L	2p	6	10
	2s	2	
K	1s	2	2

Common Pitfalls

8. TRUE or FALSE: Elements with filled shells are chemically active.

36.7 Optical Spectra and X-Ray Spectra

In a Nutshell

One result of Schrödinger's theory is that an atom possesses discrete energy levels into which its electrons can be excited. When an electron in an atom makes a transition to a lower energy level, it emits a photon of frequency $f = (E_i - E_f)/h$ and corresponding wavelength $\lambda = c/f = hc/(E_i - E_f)$. The values of the discrete wavelengths emitted by an atom create a characteristic signature of that atom. Except for hydrogen-like atoms, the Schrödinger equation is too complicated to solve analytically, so most energy levels are determined experimentally.

Not all transitions between the various energy levels of an atom can actually take place. Only those transitions that obey the following **selection rules** are allowed.

$$\Delta m = \pm 1 \text{ or } 0$$
$$\Delta \ell = \pm 1$$
$$\Delta j = \pm 1 \text{ or } 0 \text{ (but } j = 0 \rightarrow j = 0 \text{ is forbidden)}$$

Transitions involving excitations of outer (valence) electrons result in photons whose wavelengths are in or near the visible or **optical spectrum**. An energy-level diagram for sodium is shown. The first column of energy levels corresponds to $^2S_{1/2}$ states, so for these energy levels $\ell = 0$, $s = \frac{1}{2}$ and $j = \frac{1}{2}$. Except for the S states, where $\ell = 0$, all the states of sodium have two values for j, $j = \ell - \frac{1}{2}$ and $j = \ell + \frac{1}{2}$, corresponding to the two values of spin. The spin–orbit effect gives two slightly different values of energy, with the $j = \ell + \frac{1}{2}$ energy levels being at a higher energy than the $j = \ell - \frac{1}{2}$ levels. This energy difference is on the order of 0.001 eV, which is much too small to be seen on the scale used in the figure, where they appear as a single line. These states are referred to as **doublets**.

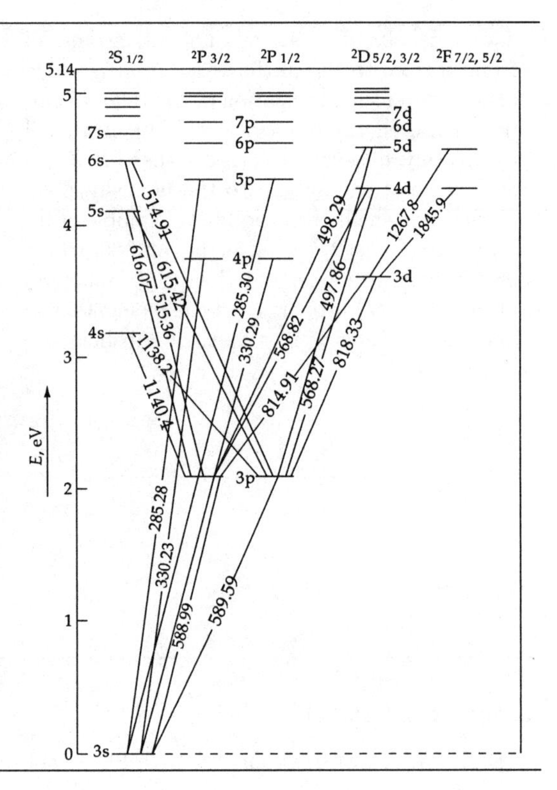

X rays, first observed by Röntgen in 1895, are high-energy photons that can be produced when a target in a cathode-ray tube is bombarded with electrons that have been accelerated through a voltage V, giving them a kinetic energy $K = eV$. The electrons striking the target lose speed rapidly, emitting photons of continuously varying wavelengths, which give rise to a continuous **bremsstrahlung spectrum** ("braking radiation" spectrum). This is the curved part of the X-ray spectrum shown. The maximum frequency that a photon can have in the bremsstrahlung spectrum is given by $hf_{\text{max}} = eV$, where eV is the kinetic energy of the bombarding electrons. This corresponds to a cutoff minimum wavelength, $\lambda_{\text{m}} = hc/E = hc/(eV)$ as shown. The cutoff wavelength λ_{m} depends only on the accelerating voltage V of the X-ray tube, and not on the material of the target.

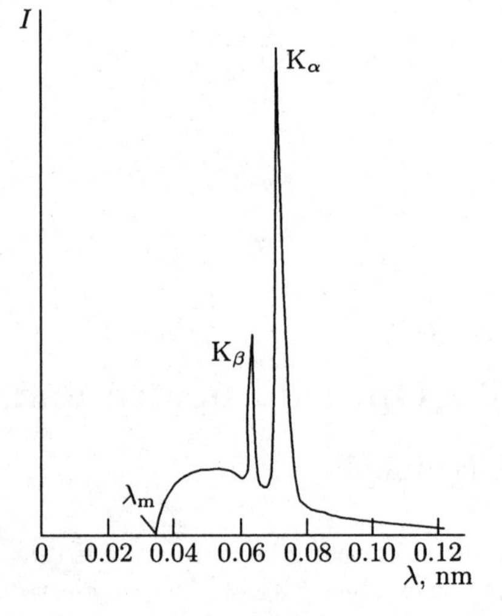

The sharp K_α and K_β spikes show the **characteristic spectrum**, which is superimposed on the continuous bremsstrahlung spectrum. The wavelengths of these peaks do depend on the target material. The characteristic spectrum arises because the bombarding electrons knock out inner-core electrons from atoms. Other electrons in the atom then make transitions from higher energy levels to the vacant energy levels, emitting the characteristic X-ray photons.

The labels of X rays arise from the letter value of the atomic shell from which the inner electron was removed. The K_α and K_β lines result from transitions from the $n = 2$ and $n = 3$ levels to the K shell ($n = 1$). Similarly, transitions to the L shell ($n = 2$) are labeled L_α and L_β. In actuality, these lines show fine structure from the splitting of the energy levels above the K shell.

In 1913 Moseley measured the wavelengths of characteristic X-ray lines for about 40 elements. For K lines, caused when one of the two electrons is knocked out of the $n = 1$ energy level, it was found that his data fit very well into the theoretical **Moseley relationship,**

$$\lambda = \frac{hc}{E_n - E_1} = \frac{hc}{(Z-1)^2(13.6\text{ eV})(1-1/n^2)}$$

which follows from a Bohr-type model in which electrons in shells outside the K shell feel the charge $+Ze$ of the nucleus and the charge $-e$ of the electron that remains in the K shell, or a net charge of $(Z-1)e$. By measuring wavelengths of characteristic X-ray spectra, Moseley was able to determine atomic numbers Z for various elements. Now X-ray spectra can be very useful in determining the elements that make up a substance under investigation.

Important Derived Results

Cutoff X-ray wavelength	$\lambda_m = \frac{hc}{E} = \frac{hc}{eV}$
Moseley relationship for X-ray wavelengths	$\lambda = \frac{hc}{E_n - E_1} = \frac{hc}{(Z-1)^2(13.6\text{ eV})(1-1/n^2)}$

Common Pitfalls

> Understand that λ_m of the bremsstrahlung part of an X-ray spectrum depends only on the accelerating voltage of the bombarding electrons and has nothing to do with the material of the target.

> Remember that the wavelengths of the X-ray photons emitted as the target electrons return to lower energy levels depend only on the structure of the target material and have nothing to do with how much energy the bombarding electrons have (as long as this energy is sufficiently large to eject the inner electrons).

9. TRUE or FALSE: Photons emitted from an X-ray tube have a maximum energy that depends on the value of the voltage that accelerates the bombarding electrons.

10. How are the wavelengths of the characteristic peaks in an X-ray spectrum affected by increasing the accelerating voltage? Explain.

They are not. Characteristic peaks come from the electrons in the material.

Try It Yourself #7

Originally, Ni (atomic weight 58.69) was listed with atomic number $Z = 27$ in the periodic table before Co (atomic weight 58.93), which was assigned $Z = 28$. Moseley measured the K_α line of Co to be 0.179 nm, and the K_α line of Ni to be 0.166 nm. Show from Moseley's data that the ordering of Ni and Co should be reversed from that originally given in the periodic table.

Picture: Substitute the wavelength into Moseley's relationship and solve for Z.

Solve:

Substitute the wavelength from Ni into Moseley's relationship and solve for Z. For K lines, $n = 2$.	$\lambda = \frac{hc}{(Z-1)^2(13.6\text{eV})(1-1/n^2)}$ $.166\text{nm} = \frac{1240\text{eV}}{(Z-1)^2(13.6\text{eV})(3/4)}$ $Z_{\text{Ni}} = 28.1$
Do the same for Co.	$.179\text{nm} = \frac{1240\text{ eV}}{(Z-1)^2(13.6\text{eV})(3/4)}$ $Z_{\text{Co}} = 27.1$

Check: We might expect exact integer values for Z, but because of experimental uncertainties, the values are close but not exact.

Try It Yourself #8

For Mo ($Z = 42$) the binding energies of the innermost core electrons are

Electron	1s	2s	2p	2p
Binding energy (keV)	20.000	2.866	2.625	2.520

Calculate the wavelengths of the fine-structure K_α lines.

Picture: From the energy differences between levels, find the wavelengths of the emitted photons consistent with the selection rule $\Delta\ell = \pm 1$.

Solve:

Find the energy difference between the first 2p state and the 1s state.	20 − 2.625 = 17.375 keV
Find the wavelength associated with this energy difference.	$E = \frac{hc}{\lambda} \rightarrow 17.375E3 = \frac{1240 eV}{\lambda}$ $\lambda \approx .0714$ nm $\lambda = 0.0714$ nm
Find the energy difference between the second 2p state and the 1s state.	20 − 2.520 = 17.48 keV
Find the wavelength associated with this energy difference.	$17.48E3 = \frac{1240 eV}{\lambda}$ $\lambda \approx .0709$ nm $\lambda = 0.0709$ nm

Check: As these are K_α lines, we should expect wavelengths much shorter than the visible spectrum, which ranges from 400 to 700 nm. The values found meet that criterion.

QUIZ

1. TRUE or FALSE: In the Bohr model, the wavelengths of a spectral series such as the Lyman series or Balmer series are calculated from the Rydberg–Ritz formula by fixing the value of n of the initial energy level.

   final

2. TRUE or FALSE: There can be only one electron with a given value of m_ℓ for a given pair of values for n and ℓ.

3. According to the Bohr model, how do the energies of the electron in hydrogen-like atoms vary with the number of protons in the nucleus?

   $E_n = \frac{-Z^2 E_0}{n^2}$ → directly proportional

4. What effect does increasing the accelerating voltage have on the X-ray spectrum for a given target?

   No effect other than increasing speed of incident photons (higher E)

5. How does the $(Z-1)e$ term in the Moseley relationship for the wavelengths of K X-ray lines arise?

   Because of the felt effects of the (+) nucleus & remaining e^- in the K shell of the atom.

6. For a given small interval Δr calculate the ratio of the probabilities of finding the electron at the two radii found in Try It Yourself #4.

$P(r) = C r^2 e^{-2Zr/a_0}$ @ $r/a_0 = .764,\ 5.24$, $Z = 1$ (Hydrogen)

$P(r)_{.764} = C r^2 e^{-1.528 Z}$ ($Z \to 1$) ; $P(r)_{5.24} = C r^2 e^{10.48 Z}$ ($Z \to 1$)

$P(r)_{.764} = C_{200} (2 - .764) e^{-.764}$, $P(r)_{5.24} = C_{200} (2 - 5.24) e^{-5.24}$

7. When an electron in a galactic hydrogen atom "flips" its spin from aligned to anti-aligned with the spin of the proton in an atom, microwave radiation of wavelength 21.0 cm is emitted, which is used in radioastronomy to map the galaxy. Determine the magnetic field felt by the electron in a galactic hydrogen atom.

Chapter 37

Molecules

37.1 Bonding

In a Nutshell

The strongest interactive forces among atoms, created by sharing one or more electrons, are attractive forces called **bonds**, which bind two or more atoms together into a molecule.

The **ionic bond** is the force that holds the atoms of most salt crystals together. In the simplest salts an alkali metal from the first column of the period table (Li, Na, K, ...) bonds with a halogen from the next-to-last column (Fl, Cl, Br, ...). We examine sodium chloride (NaCl) below to illustrate ionic bonding.

Sodium ($Z = 11$) has an atomic configuration $1s^2 2s^2 2p^6 3s^1$: a closed inert core plus an outer 3s electron. The last 3s electron is weakly bound to the inner core. A bond is referred to as "weak" or "strong" according to the energy required to break it. The energy needed to remove one electron from a Na atom to form a positive Na^+ ion, that is the **ionization energy** I for Na, is only 5.14 eV. The ionization process is indicated by the reaction $Na + I \rightarrow Na^+ + e^-$.

Chlorine ($Z = 17$) has an atomic configuration $1s^2 2s^2 2p^6 3s^2 3p^5$. Adding one more 3p electron would form a closed inert core. The **electron affinity** A measures the energy released when an atom gains an electron; for Cl $A = 3.62$ eV is required to add an electron to a Cl^- ion to form a Cl atom, as indicated by the reaction $Cl^- + A \rightarrow Cl + e^-$.

The formation of a NaCl molecule from the neutral atoms can be regarded as occurring in two steps. First, an electron is removed from Na and transferred to Cl to form the ions Na^+ and Cl^-; then the ions are combined to form NaCl. The energy required for the first step is obtained by adding the ionization energy reaction to the inverse of the electron affinity reaction to obtain $Na + Cl + (I - A) \rightarrow Na^+ + Cl^-$. Thus the electron transfer from Na to Cl requires energy $I - A = 5.14 \text{ eV} - 3.62 \text{ eV} = 1.52 \text{ eV}$.

The Na^+ and Cl^- ions are mutually attracted by a Coulomb force: the corresponding electrostatic potential energy $-ke^2/r$ decreases as the distance r between them diminishes. When the distance between the two ions becomes very small, the Coulomb attraction between them is overwhelmed by exclusion-principle repulsion, which arises because the wave functions of the core electrons of the two ions overlap at small distances. Because, by the exclusion principle, no two electrons in a system can occupy the same state, some of the electrons in the overlap region must go into higher energy states. To accomplish this requires work against the repulsive forces, thereby increasing the potential energy of the ionic pair.

The total potential energy of the Na^+ and Cl^- ions is the sum of their negative electrostatic potential energy and their positive repulsive potential energy. To express this relationship quantitatively, we first must choose a value for the potential energy when the ions are infinitely separated and there is no overlap region. We let this energy correspond to the energy of neutral Na and Cl atoms; as a result, the energy of a pair of Na^+ and Cl^- ions at infinite separation is 1.52 eV, and the total potential energy as a function of the separation r is

$$\begin{aligned} U(r) &= U_{\text{electrostatic}} + U_{\text{repulsive}} + 1.52 \text{ eV} \\ &= -ke^2/r + U_{\text{repulsive}} + 1.52 \text{ eV} \end{aligned}$$

The $U(r)$ versus r curve for NaCl is shown below. The **equilibrium separation** occurs at the minimum of the $U(r)$ curve, which is found experimentally to be at a separation $r_0 = 0.236$ nm, corresponding to $U(r_0) = -4.27$ eV. The **dissociation energy** D, which is the energy needed to break up the molecule into neutral Na and Cl atoms, is then $E = 4.27$ eV.

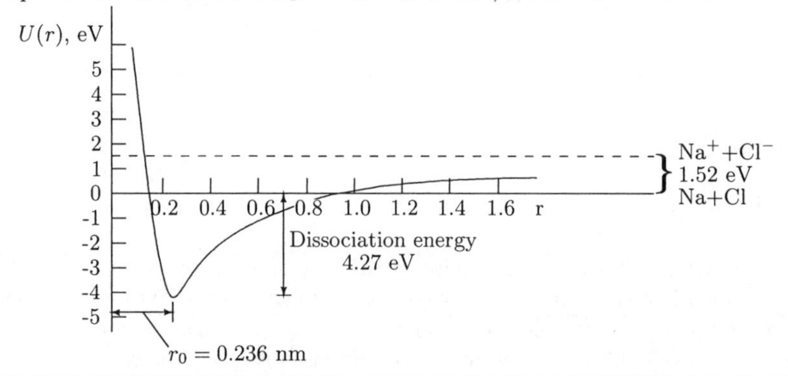

The **covalent bond** arises from quantum-mechanical effects that bind molecules formed from identical or similar atoms, such as hydrogen (H_2), nitrogen (N_2), carbon monoxide (CO), hydrogen fluoride (HF), and hydrogen chloride (HCl). To illustrate the quantum-mechanical nature of covalent bonding, we will examine the hydrogen molecule H_2, which is formed from two hydrogen atoms whose individual wave functions were described in Chapter 36.

When two hydrogen atoms combine to form H_2, the two electron spins are either antiparallel or parallel. When the electrons are antiparallel, the total wave function for the H_2 molecule, composed of the wave functions of the individual atoms, is symmetrical, and is denoted by ψ_S as shown in Figure (a). When the electrons are parallel, the total wave function ψ_A is antisymmetric, as shown in Figure (b). The square of these wave functions, shown in Figures (c) and (d), gives the probability distribution for finding the two electrons in the space around the two protons. Notice in Figure (c) that for antiparallel spins the probability is large for finding the electrons between the two protons, whereas in Figure (d) the probability is small for finding electrons with parallel spins between the two protons. The "probability cloud" represents the probable locations of the electrons. The negative cloud of electrons with antiparallel spins attracts the protons, producing a bonding force between the protons, whereas electrons with parallel spins create no negative charge between the protons and thus no bonding.

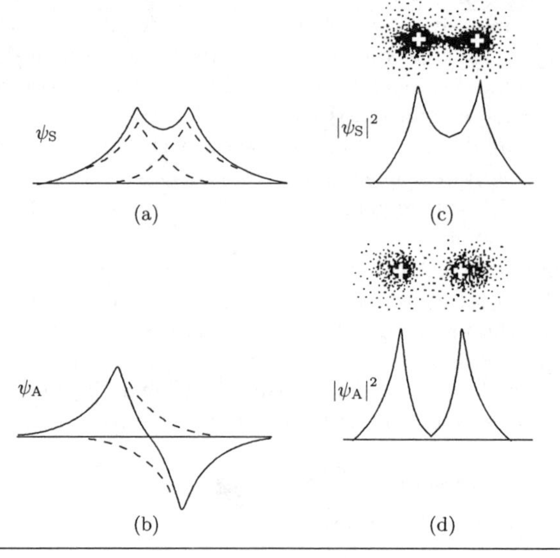

The figure below illustrates these two cases in a different way. The total electrostatic potential energy consists of a positive potential energy that corresponds to repulsive forces between the two electrons and a negative potential energy that corresponds to the attractive force between the electrons and protons. As the distance increases, the antisymmetric U_{A} potential energy curve, corresponding to parallel spins, decreases steadily but never becomes negative, so there is no equilibrium position—that is, there is no bonding. On the other hand, the symmetric U_{S} potential energy curve, corresponding to antiparallel spins, has a minimum at which is the equilibrium separation of the two protons. The figure also shows that the dissociation energy required to separate the hydrogen molecule into two neutral atoms is 4.52 eV.

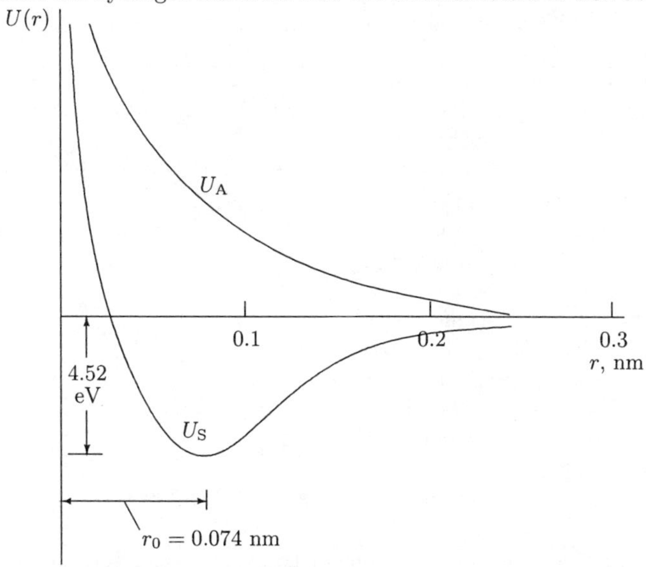

The difference in the physical basis for ionic and covalent bonds is that an ionic bond is created when an electron is effectively transferred from one atom to the other, causing the two oppositely charged ions to attract each other. In a covalent bond the protons share the electrons, causing the two positively charged nuclei to be attracted to the cloud of negatively charged electrons between them.

The bond between two identical atoms is totally covalent. In molecules formed from two different atoms, the bonding is usually a mixture of ionic and covalent bonds. In NaCl there is some covalent bonding, and in CO there is some ionic bonding. The degree (or proportion) of ionic and covalent bonding in a molecule can be determined from the magnitude of the molecule's **electric dipole moment**. The electric dipole moment $\vec{p}$ of two equal and opposite charges Q separated by a distance d has a magnitude $p = Qd$ and a direction that points from the negative to the positive charge.

The distance between two oppositely charged ions in a molecule with ionic bonding is relatively large, resulting in a relatively large dipole moment. The negative charge of the two electrons in a molecule with covalent bonding is located mostly between the two positively charged nuclei, producing two dipole moments that point in opposite directions and therefore tend to cancel, resulting in a relatively small overall dipole moment.

The **van der Waals bond** arises from weak attractive forces between the dipole moments of molecules. It is the van der Waals force that bonds nearly all substances in the liquid and solid states. Some molecules, such as H_2O, have permanent dipole moments that attract each other to produce bonding in the liquid and solid states, as shown. Other molecules, which have zero dipole moments, can induce dipole moments in each other, causing a van der Waals force between them.

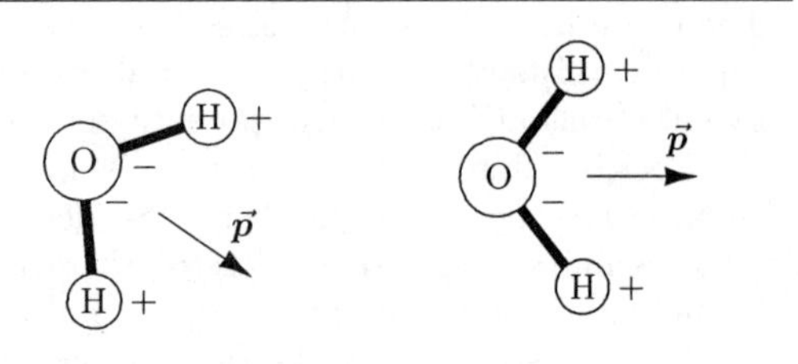

In a **hydrogen bond**, a hydrogen atom is shared between two other atoms. As an example, consider the hydrogen bonding between two water molecules, as shown. The electron of a hydrogen atom H in one H_2O molecule is attracted to (spends more time in the vicinity of) an oxygen atom O of an adjacent H_2O molecule. Effectively, the H atom becomes positively charged and the O atom becomes negatively charged. The resulting Coulomb attraction constitutes the hydrogen bond between the two water molecules. Hydrogen bonds often link groups of molecules and play an important role in enabling giant biological molecules, such as the DNA molecule, to retain their shape.

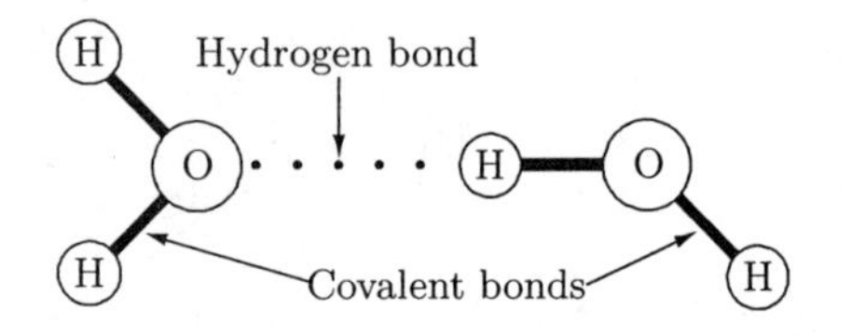

In a metallic bond many electrons—roughly one or two valence electrons per atom—move quite freely as an electron "gas" throughout the metal. The weak attractive force between the positive ions and the electron gas provides the bonding mechanism that holds the ions together.

Common Pitfalls

> It is easy to confuse ionic and covalent bonds. In an ionic bond, an electron is effectively transferred from one atom to another, producing two oppositely charged ions that attract each other. In a covalent bond, the valence electrons are shared by two nuclei. The two positively charged nuclei are attracted to the negatively charged electron cloud between them.

1. TRUE or FALSE: The bonding between two dissimilar atoms is always either totally ionic or totally covalent.

2. Why don't the two oppositely charged ions in an ionically bound molecule continue to attract each other to very small separations?

Try It Yourself #1

The measured dissociation energy of LiF is 5.95 eV, the ionization energy of Li is 5.39 eV, and the electron affinity of F is 3.45 eV. If the repulsive energy at the equilibrium separation of LiF is 1.34 eV, estimate the dipole moment of LiF.

Picture: You can find r_0 from the repulsive Coulomb potential energy. Use that to find the dipole moment of LiF.

Solve:

Write an *algebraic* expression for the Coulomb potential energy.	$U(r) = -\frac{ke^2}{r} + U_{rep} + U_{e^-static}$
From the Coulomb energy, determine an algebraic expression for r_0.	
Write a generic *algebraic* expression for the magnitude of the dipole moment.	
Substitute the value for q and your expression for r_0 from above to find the dipole moment of LiF.	$p = 2.49 \times 10^{-29}\ \text{C} \cdot \text{m}$

Check: The units are correct. This seems like a pretty small number, but it should be. The charge involved is on the order of 10^{-19} C, and nuclear separations are approximately 10^{-10} m, so this value of dipole moment is reasonable.

Taking It Further: How would you expect the dipole moment of MgO to compare to that calculated here for LiF?

Try It Yourself #2

The equilibrium separation of the two protons in an H_2 molecule is 0.074 nm, and the measured binding energy is 4.50 eV. If only Coulomb interactions are taking place, how much negative charge has to be placed at the midpoint between the two nuclei to account for the measured binding energy?

Picture: The binding energy is the difference between the positive Coulomb energy between the two protons at equilibrium separation and the negative Coulomb energy between the protons and the unknown negative charge midway between them.

Solve:

Determine the positive Coulomb energy between the two protons at their equilibrium separation.	
Write an *algebraic* expression for the negative Coulomb energy between the two protons and the unknown negative charge between them.	
Solve for the unknown negative charge by equating the difference of the energies in the first two steps to the binding energy.	$q = -0.308e$

Check: This charge is on the order of the charge on a single electron, so from that standpoint, it seems to make sense. However, charges come only in units of e, so something else must be going on.

Taking It Further: Why is this so much smaller than $-2e$, the charge from the two electrons in an H_2 atom?

37.2 Polyatomic Molecules*

In a Nutshell

The structure of polyatomic molecules—molecules with more than two atoms—follows from the principles of quantum mechanics applied to the bonding between individual atoms. It is primarily the covalent and hydrogen bonds that provide the bonding mechanism of polyatomic molecules.

37.3 Energy Levels and Spectra of Diatomic Molecules

In a Nutshell

The electrons in molecules, as in atoms, can be excited into higher energy states. In addition to these electronic excitations, the molecule as a whole rotates and vibrates, and both these motions are quantized and have energies much lower than the energies of the electronic states. In the following we will restrict ourselves to the simplest case of diatomic (two-atom) molecules.

A rotating diatomic molecule can be pictured classically as a dumbbell with unequal masses rotating in a plane about its center of mass, as shown. The kinetic energy, with respect to its center of mass, of this molecule, which has moment of inertia I about its center of mass, and rotates at an angular frequency ω, is $K = \frac{1}{2}I\omega^2$, which can be rewritten in terms of the angular momentum $L = I\omega$ as $K = L^2/2I$.

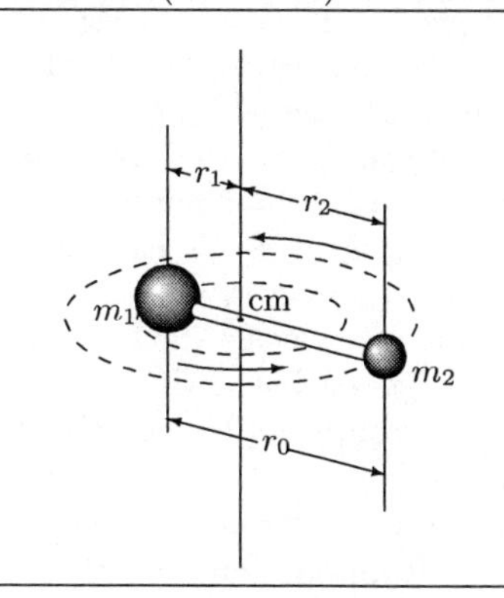

Solving the Schrödinger equation shows that the rotational angular momentum is quantized according to $L^2 = \ell(\ell+1)\hbar^2$, where ℓ is the **rotational quantum number** and has values given by $\ell = 0, 1, 2, 3, \ldots$. This clearly results in quantized rotational energy levels with energies given by $E = \ell(\ell+1)\hbar^2/(2I) = \ell(\ell+1)E_{0r}$, for $\ell = 0, 1, 2, 3, \ldots$. $E_{0r} = \hbar^2/(2I)$ is the **characteristic rotational energy** for a given molecule of moment of inertia I. A rotational energy-level diagram is shown. Only transitions that satisfy the selection rule $\Delta\ell = \pm 1$ are allowed.

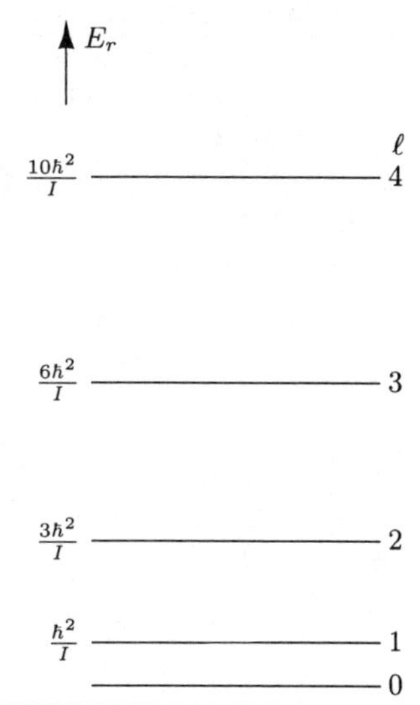

If a diatomic molecule is thought of as a dumbbell with masses m_1 and m_2 located at distances r_1 and r_2 from the center of mass, as shown above, the moment of inertia of the dumbbell $I = m_1r_1^2 + m_2r_2^2$ can be rewritten in terms of the **reduced mass** $\mu = m_1m_2/(m_1 + m_2)$ as $I = \mu r_0^2$, where $r_0 = r_1 + r_2$.

*Optional material

In addition to rotating, a diatomic molecule can also vibrate in a manner similar to two masses vibrating at the ends of a spring as shown.

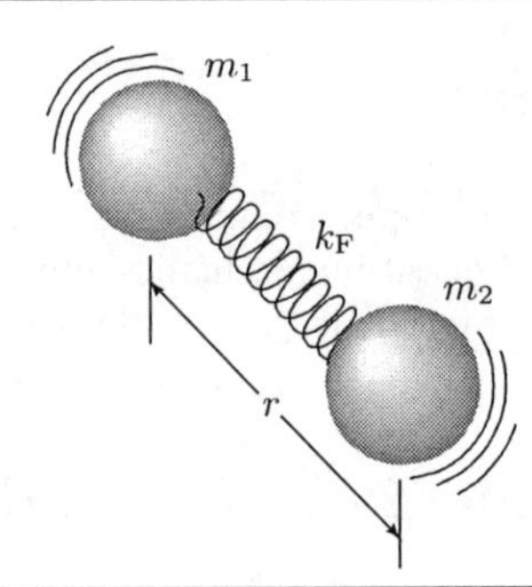

From the Schrödinger equation for a simple harmonic oscillator, it is found that the vibrational energies are quantized according to $E_\nu = (\nu + \frac{1}{2})hf$ for $\nu = 0, 1, 2, \ldots$, where ν is the **vibrational quantum number** and f is the frequency of vibration. An energy-level diagram is shown. The energy levels are equally spaced by an amount $\Delta E = hf$. Only transitions that satisfy the selection rule $\Delta\nu = 1$ are allowed for transitions within the same electronic state, so a photon emitted by a transition between energy levels has the frequency f. The lowest energy level is not zero but is $\frac{1}{2}hf$. Thus, a molecule in its lowest energy state is not at rest but is vibrating with some **zero-point minimum energy** about its equilibrium position.

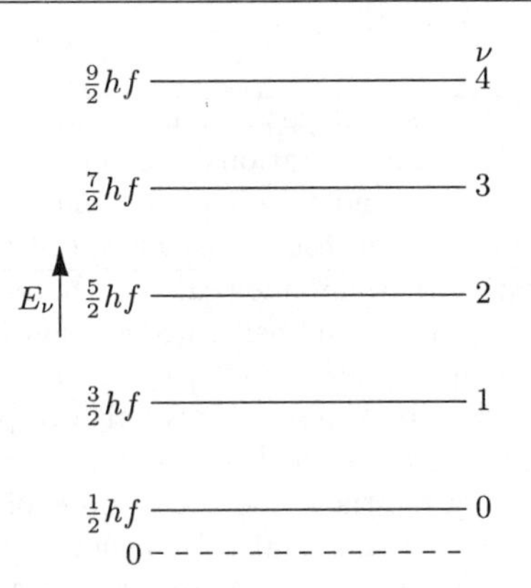

The **effective force constant** k_F of a diatomic molecule can be determined by measuring the vibrational frequency and using the relation for two masses vibrating on a spring: $f = \frac{1}{2\pi}\sqrt{k_F/\mu}$, where μ is the reduced mass described earlier.

The spacing between energy levels of electronic states is much greater than the spacing between vibrational energy levels, which in turn is much greater than the spacing between rotational energy levels. As a consequence, we can regard vibrational energy levels as being built up from each electronic energy level, and the smaller rotational energy levels can be treated as being built on each vibrational energy level. The situation is illustrated below.

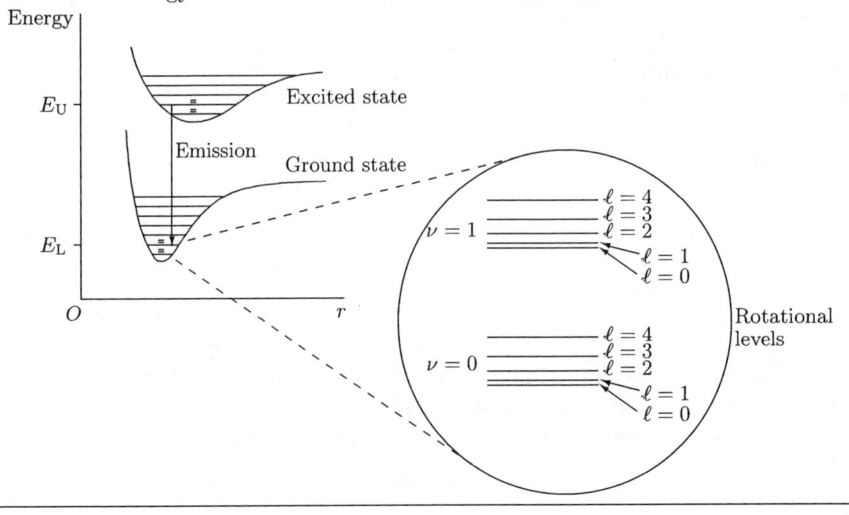

A transition from a vibrational energy level E_U of an upper electronic state to a vibrational energy level E_L of a lower electronic state as shown above results in the emission of a photon, whose wavelength is given by $hc/\lambda = E_U - E_L$. The selection rule $\Delta\nu = \pm 1$ does not hold in this situation.

The observed **emission spectrum** appears as vibrational bands (one for each transition $\Delta\nu = \nu_U - \nu_L$) rather than isolated lines. With higher resolution, each band shows a fine structure: a collection of spectral lines. Each line in the fine structure arises from a transition between the various rotational energy levels that exist at each vibrational energy level.

Absorption spectra show dark lines where photons have been removed from the incident beam as a result of exciting molecules in the ground electronic state from lower to higher vibrational and rotational energy levels. The predominant transition in the infrared is from the $\nu = 0$ to the $\nu = 1$ vibrational energy level. In this transition the atom is excited from various rotational energy levels at the vibrational level $\nu = 0$ to other rotational levels at the higher vibrational level $\nu = 1$, subject to the selection rule $\Delta\nu = \pm 1$, as shown below.

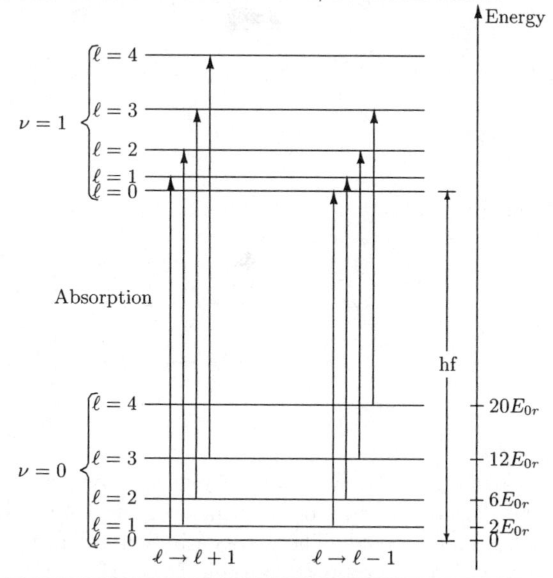

The frequencies of the vibrational-rotational absorption lines fall into two groups,

$$f_{\ell \to \ell+1} = \frac{\Delta E_{\ell \to \ell+1}}{h} = f + \frac{2(\ell+1)E_{0r}}{h}, \ell = 0, 1, 2, \ldots$$
$$f_{\ell \to \ell-1} = \frac{\Delta E_{\ell \to \ell-1}}{h} = f - \frac{2\ell E_{0r}}{h}, \ell = 0, 1, 2, \ldots$$

which result in the lines shown below. The center of the gap occurs at a photon energy of hf, and the separation between the absorption energies of adjacent lines in the two groups on each side of the gap is $2E_{0r}$.

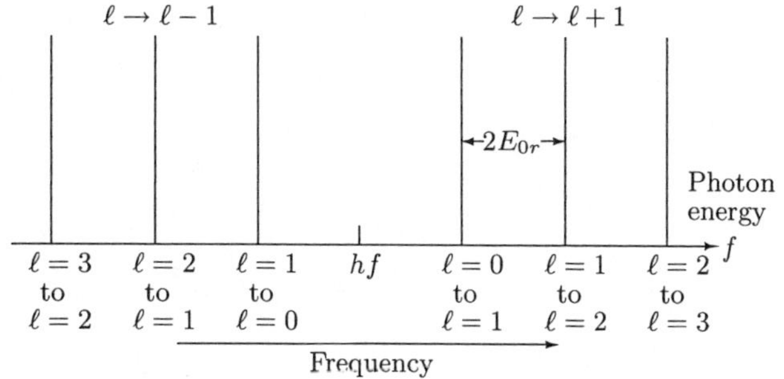

Important Derived Results

Classical kinetic energy of rotation	$E = \frac{1}{2}I\omega^2 = \frac{L^2}{2I}$
Quantized angular momentum	$L^2 = \ell(\ell+1)\hbar^2$, for $\ell = 0, 1, 2, \ldots$
Quantized rotational energies	$E = \frac{\ell(\ell+1)\hbar^2}{2I} = \ell(\ell+1)E_{0r}$, for $\ell = 0, 1, 2, \ldots$
Moment of inertia of a dumbbell	$I = m_1 r_1^2 + m_2 r_2^2 = \mu r_0^2$
Reduced mass	$\mu = \frac{m_1 m_2}{m_1 + m_2}$
Quantized vibrational energies	$E = \left(\nu + \frac{1}{2}\right)hf$, for $\nu = 0.1, 2, \ldots$
Angular momentum selection rule	$\Delta\ell = \pm 1$
Vibrational state selection rule	$\Delta\nu = \pm 1$
Relationship between frequency and effective force constant	$f = \frac{1}{2\pi}\sqrt{\frac{k_f}{\mu}}$
Frequencies of absorption lines	$f_{\ell\to\ell+1} = \frac{\Delta E_{\ell\to\ell+1}}{h} = f + \frac{2(\ell+1)E_{0r}}{h}, \ell = 0, 1, 2, \ldots$ $f_{\ell\to\ell-1} = \frac{\Delta E_{\ell\to\ell-1}}{h} = f - \frac{2\ell E_{0r}}{h}, \ell = 0, 1, 2, \ldots$

Common Pitfalls

> Do not confuse rotational energies that are specified by the quantum number ℓ with the orbital angular momenta of the electrons in an atom of a molecule, which are specified by a quantum number designated by the same letter ℓ. These are entirely different physical quantities. With rotational energy levels, the quantization condition $L^2 = \ell(\ell+1)\hbar^2$ refers to the angular momentum of the molecule as a whole rotating about its center of mass. This has nothing to do with the orbital angular momenta of the electrons in the molecule.

> When dealing with combined rotational and vibrational excitations, do not mix up the different types of energy levels. The smaller-spaced rotational energy levels are built on each larger-spaced vibrational energy level and not the other way around.

3. TRUE or FALSE: The lowest vibrational energy level corresponds to zero energy.

4. How can you experimentally measure the moment of inertia of a diatomic molecule?

Find ℓ. Then plug into $E = \frac{\ell(\ell+1)\hbar^2}{2I}$

Try It Yourself #3

The characteristic rotational energy of a CO molecule is 2.39×10^{-4} eV.What is the interatomic spacing of the molecule? What is the frequency of radiation corresponding to a transition from the $\ell = 0$ to $\ell = 1$ rotational state of the molecule?

Picture: The moment of inertia can be expressed in terms of the reduced mass and the interatomic spacing. The frequency is proportional to the energy difference between the rotational states.

Solve:

Find the moment of inertia from the characteristic rotational energy.	
Express the moment of inertia in terms of the reduced mass and interatomic spacing to solve for the spacing.	$r_0 = 0.113$ nm
Determine the energy of the $\ell = 0$ state.	
Determine the energy of the $\ell = 1$ state.	
Determine the frequency associated with this energy difference.	$f = 1.16 \times 10^{11}$ Hz

Check: The interatomic spacing of less than a nanometer is reasonable. This frequency corresponds to the microwave region of radiation, whose relatively low energies should be expected for rotational transitions.

Try It Yourself #4

The center of the gap of the vibrational-rotational absorption spectrum of HCl gas occurs at 3465 nm. What are the vibrational frequency and the effective force constant of a HCl molecule?

Picture: The wavelength is inversely related to the frequency of the center of the gap. Use the reduced mass and the vibrational frequency in the expression for the effective force constant.

Solve:

Relate the wavelength of the center of the gap to the frequency and speed of light.	$f = 8.66 \times 10^{13}$ Hz
Look up the masses of H and Cl.	
Determine the reduced mass of the HCl molecule.	
Use the reduced mass and the frequency found in the first step to find the effective force constant.	$k_F = 478$ N/m

Check: The characteristic frequency is in the infrared region, as expected. Force constants on the order of hundreds of N/m are also to be expected.

QUIZ

1. TRUE or FALSE: The energy difference between adjacent rotational energy levels is larger than the energy difference between adjacent vibrational energy levels.

2. TRUE or FALSE: Wide bands are seen in the emission spectra of molecules because the vibrational energy levels are not well defined.

3. Which has the larger dipole moment, a molecule with ionic bonding or a molecule with covalent bonding? Why?

   Ionic bonds do not have dipoles that nearly cancel like covalent bonds.

4. What is zero-point vibrational energy?

   $E = (0 + 1/2)\,hf$

5. Briefly explain the mechanism of the van der Waals bond.

   Van der Waals bond = attraction b/t (induced or permenant) dipoles

6. The following quantities can be measured in an ionically bonded molecule: r_0, the equilibrium separation; I, the ionization energy of the alkali atom; A, the electron affinity of the halogen atom; and R, the energy of repulsion at the equilibrium separation. Obtain an expression for the dissociation energy D of the molecule in terms of these quantities and other physical constants.

$$D = \left(\frac{ke^2}{r_0}\right) - R + A - I$$

7. The vibrational-rotational absorption spectrum for a diatomic molecule is shown on page 73 with the photon frequencies of the peaks separated by 1.2×10^{12} Hz and the center of the gap at 110×10^{12} Hz. Find the lowest vibrational energy of the molecule that produced this absorption spectrum.

Lowest E @ $v = 0$

$\therefore \; f_{\ell \to \ell-1} = f - \frac{2\ell E_{0r}}{h}$, $\ell = 1$

$f_\ell \to 110\text{E}12\text{Hz} - \frac{(1)(2E_{0r})}{h}$ → $h(1.2\text{E}12)$

$f_\ell \to 102\text{E}12\text{Hz}$

$E = (6.626\text{E}{-34})(102\text{E}12\text{Hz}) = .421$

Chapter 38

Solids

38.1 The Structure of Solids

In a Nutshell

By cooling a liquid sufficiently, we can form one of two kinds of solids: amorphous and crystalline. The molecules of an **amorphous solid**, which is usually formed by rapid cooling, are not arranged in regular arrays but instead are oriented randomly. An amorphous solid does not have a well-defined melting point; it just softens as its temperature increases. Glass is a common amorphous solid.

The molecules of a **crystalline solid**, which has a well-defined melting temperature, are arranged in regular, symmetrical arrays over relatively large regions of the crystal. A single unit structure, called a **unit cell**, repeats regularly throughout the crystal. The makeup of the unit cell depends on the type of bonding between the atoms, ions, or molecules of the crystal, as discussed in Chapter 37.

The structure of a sodium chloride (NaCl) ionic crystal is shown here. In a NaCl crystal, each Na^+ ion has six Cl^- ions as its nearest neighbors, and vice versa; this structure is called a **face-centered-cubic crystal**.

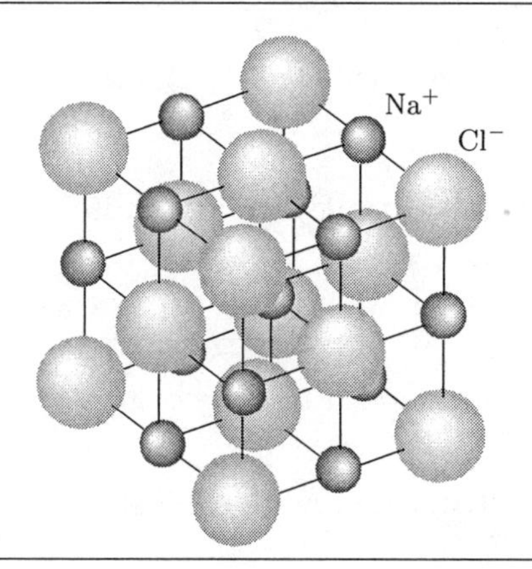

In a **body-centered-cubic crystal**, each atom has eight nearest neighbor ions of the opposite charge; the figure shows the structure of the body-centered-cubic crystal CsCl.

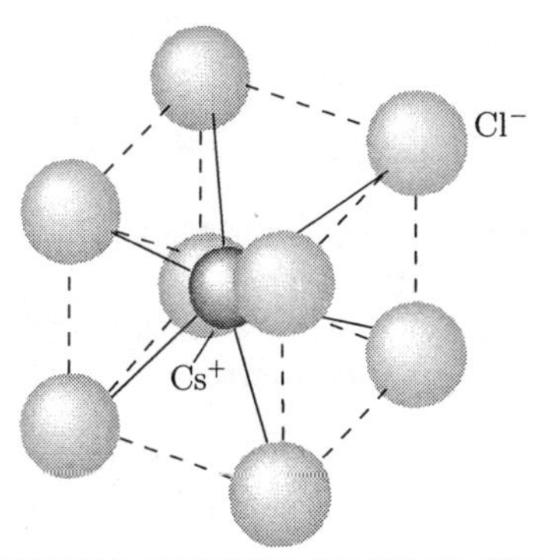

The potential energy of a given ion in a crystal shared with another ion in the crystal is $U = \pm ke^2/R$, where R is the distance from the given ion to the other ion, and the sign depends on the charges of the two ions. Adding the potential energies due to the various ions in a crystal results in the following expression for the net attractive potential energy of an ion in a crystal: $U_{\text{att}} = -\alpha ke^2/r$, where k is the Coulomb constant, r is the distance between adjacent ions and the **Madelung constant** α depends on the geometry of the structure of a unit cell. For face-centered-cubic structures $\alpha = 1.7476$ and for body-centered-cubic structures $\alpha = 1.7627$.

In addition to the Coulomb attractive potential energy between ions in an ionic crystal, there also is an exclusion-principle repulsion potential energy. This is given by the empirical expression $U_{\text{rep}} = A/r^n$, where A and n are constants. Thus the **total potential energy** of one ion due to the presence of the other ions in a crystal is

$$U(r) = -\alpha\frac{ke^2}{r} + \frac{A}{r^n}$$

The **equilibrium separation** r_0 is the value of r at which $U(r)$ is minimum. To find A in terms of r_0 set $dU/dr = 0$ and $r = r_0$ to obtain $A = \alpha ke^2 r_0^{n-1}/n$, which enables us to write the total potential energy as

$$U(r) = -\alpha\frac{ke^2}{r_0}\left[\frac{r_0}{r} - \frac{1}{n}\left(\frac{r_0}{r}\right)^n\right]$$

The **dissociation energy** (the energy needed to break up the crystal into atoms) equals the magnitude of $U(r)$ at $r = r_0$:

$$\text{Dissociation energy} = |U(r_0)| = \frac{\alpha ke^2}{r_0}\left(1 - \frac{1}{n}\right)$$

Physical Quantities and Their Units

Madelung constant for face-centered cubic structures $\alpha = 1.7476$

Madelung constant for body-centered cubic structures $\alpha = 1.7627$

Important Derived Results

Dissociation energy $\quad$ $\text{Dissociation energy} = |U(r_0)| = \frac{\alpha ke^2}{r_0}\left(1 - \frac{1}{n}\right)$

Common Pitfalls

> Understand how the Madelung constant is related to the geometry of a crystal and why it is of the order of 1.8.

> Do not confuse the structures of face-centered-cubic crystals, body-centered-cubic crystals, and hexagonal close-packed crystals.

1. TRUE or FALSE: In a crystal that has a face-centered-cubic structure, each positive ion has six nearest negative-ion neighbors, and each negative ion has six nearest positive-ion neighbors.

2. Explain how the Madelung constant originates.

Depends on geometry ∘ structure ∘ unit cell.

Try It Yourself #1

LiCl crystals have a density of 2.07 g/cm^3. Knowing that the crystal structure is face-centered cubic and that the dissociation energy for one pair of ions is 4.86 eV, determine the value of the exponent n used in describing the exclusion-principle repulsion.

Picture: The equilibrium separation distance can be determined from the density and molar mass of the crystal. The dissociation energy must be equal to the potential energy when the ions are at their equilibrium separation. This can be used to solve for n.

Solve:

Assume each ion occupies a cubic volume of side length r_0. The volume of one mole of LiCl equals the number of ions multiplied by the volume per ion. Write an *algebraic* expression for the volume of one mole of LiCl.	
Algebraically relate the density to the molar mass and the volume calculated above.	
Solve for the equilibrium separation distance.	
Set the dissociation energy equal to the total potential energy at the equilibrium separation distance.	
Solve for n.	$n = 0.668$

Check: The equilibrium separation distance of less than 1 nm is expected for solids.

38.2 A Microscopic Picture of Conduction

In a Nutshell

In a classical microscopic picture of a conductor, valence electrons are not bound to particular atoms of the conductor. Instead, electrons move almost freely like a gas throughout the entire conductor. In some given volume V, a large number N of electrons randomly collide with the fixed lattice ions, as shown. Classical Newtonian theory and the classical Maxwell-Boltzmann distribution can be used to calculate such properties as resistivity, mean free path, and heat conduction.

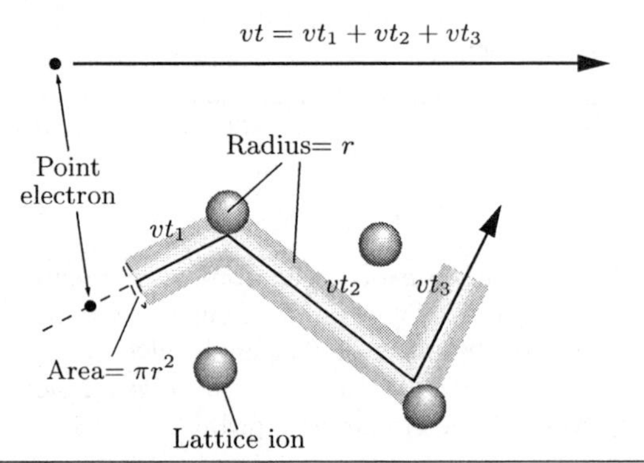

Ohm's law for a current-carrying wire segment states that the voltage V across the wire is related to the current I through the wire by $V = IR$, where the resistance R is related to the resistivity ρ by $R = \rho L/A$, with L the length and A the cross-sectional area of the wire segment. When a voltage is applied across a conductor, the resulting electric field inside the conductor causes the electrons to flow, thereby producing a current. As the electrons are accelerated through the wire under the influence of the applied electric field, they collide with the fixed ions in the wire, which slows them down. The **drift velocity** $\vec{v}_\text{d}$ of the electrons, which is superimposed on top of the random velocity of their thermal motion, is related to the electric field $\vec{E}$ and resistivity by $\vec{v}_\text{d} = \vec{E}/(\rho n_\text{e} e)$, where $n_\text{e} = N/V$ is the number of electrons per unit volume.

It is found that the drift velocity is many orders of magnitude ($\sim 10^{-9}$) smaller than the random thermal velocity. Because of this, as we consider thermal motion of electrons in the following, we are justified in neglecting the extra motion of electrons produced by an applied electric field.

If τ is the average time between collisions for a given electron in thermal motion, called the **collision time**, the resistivity can be written as $\rho = m_\text{e}/(n_\text{e}e^2\tau)$.

If v_av is the mean thermal speed of an electron between collisions, the average distance the electron travels between collisions, called the **mean free path** λ, is $\lambda = v_\text{av}\tau$, allowing the resistivity to be written as $\rho = m_\text{e}v_\text{av}/(n_\text{e}e^2\lambda)$.

The classical picture of a point electron colliding with fixed ions does not give a resistivity that agrees with experimental results. Classically, according to the Maxwell-Boltzmann distribution, an electron in an electron gas has a root mean square (rms) speed v_rms which is slightly greater than the mean speed v_av given by $v_\text{rms} = \sqrt{3kT/m_\text{e}}$, where T is the absolute temperature of the electron gas. If this expression for v_rms is set equal to v_av in the above expression for resistivity, it is seen that classical ideas predict that the resistivity will vary with temperature according to $\rho \propto \sqrt{T}$. Experimentally, though, ρ is found to vary linearly with temperature T. Further, at $T = 300$ K the classical calculation for resistivity based on the Maxwell-Boltzmann distribution gives a value for resistivity that is about six times greater than what is measured experimentally. Finally, the classical model does not explain why some materials are conductors, some are insulators, and others semiconductors.

Important Derived Results

Relation of resistance to resistivity	$R = \frac{\rho L}{A}$
Relation of resistivity to electric field and drift velocity	$\vec{v}_{\mathrm{d}} = \frac{\vec{E}}{n_{\mathrm{e}} e \rho}$
Relation of resistivity to collision time	$\rho = \frac{m_{\mathrm{e}}}{n_{\mathrm{e}} e^2 \tau}$
Relation of mean free path to collision time	$\lambda = v_{\mathrm{av}} \tau$
Relation of resistivity to mean free path and mean speed	$\rho = \frac{m_{\mathrm{e}} v_{\mathrm{av}}}{n_{\mathrm{e}} e^2 \lambda}$
Root mean square speed according to classical Maxwell-Boltzmann theory	$v_{\mathrm{rms}} = \sqrt{\frac{3kT}{m_{\mathrm{e}}}}$

Common Pitfalls

> Remember the differences between the drift velocity of electrons and their random thermal mean velocity.

3. TRUE or FALSE: In a conductor, the mean speed and root mean square speed of electrons are comparable in magnitude.

4. Explain the difference between mean velocity and drift velocity.

Drift velocity → constant collision in wire, yet steady progression (relatively slow).
Mean velocity → due to thermal motion.
* Drift velocity is in addition to mean velocity when $\vec{E}$ field is established.

Try It Yourself #2

Find the drift speed of electrons when an electric field of 9.00×10^{-3} V/m is established in a copper wire at 20°C = 293 K. Take the mean free path equal to 0.400 nm, the electron density equal to 8.47×10^{22} electrons/cm^3, and the resistivity equal to $1.70 \times 10^{-8}\ \Omega \cdot \mathrm{m}$.

Picture: Use the expression that relates the resistivity to the electric field and the drift speed.

Solve:

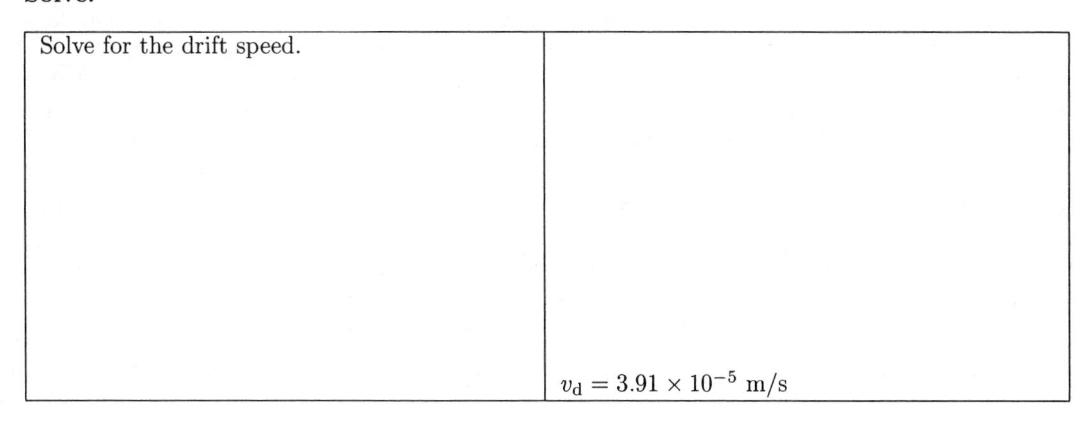

Solve for the drift speed.	$v_{\mathrm{d}} = 3.91 \times 10^{-5}$ m/s

Check: This speed is on the order of what we have been told to expect for drift speeds.

Taking It Further: What is the mean speed of the electrons in the copper wire?

38.3 Free Electrons in a Solid

In a Nutshell

The classical model of electron conduction described above fails because electrons are not Newtonian point particles following Maxwell–Boltzmann statistics having an average kinetic energy equal to $\frac{3}{2}kT$, with zero kinetic energy at $T = 0$. Instead, electrons have wave properties and obey statistics that are described by a quantum-mechanical Fermi–Dirac distribution, obeying the Pauli exclusion principle, which will be described below.

To get a feeling for the wave properties of electrons, refer to Chapter 35 where a particle of mass m is confined to a finite region of space such as a one-dimensional box of length L. For such a particle, we found that the allowed energies could take on only the quantized values $E_n = n^2h^2/(8mL^2) = n^2E_1$, for $n = 1, 2, 3, \ldots$. Here n is called the **spatial quantum number**, and $E_1 = h^2/(8mL^2)$ is the ground state energy. For a given quantum number n, the wave function for the particle is given by $\psi_n = \sqrt{2/L}\sin(n\pi x/L)$, and $\lambda_n = 2L/n$ is the wavelength associated with the particle.

The electrons in a metal cannot have any arbitrary energy E_n because they must satisfy the **Pauli exclusion principle**: "No two electrons in a system can be in the same quantum state; that is, they cannot have the same set of values for their quantum numbers." Electrons are "spin one-half" particles—that is, their spin quantum number m_s can have two possible values: $+\frac{1}{2}$ or $-\frac{1}{2}$. This means that there can be at most two electrons having a given value of the spatial quantum numbers n, ℓ, and m_ℓ. Most of our problems will be one-dimensional, so we will need to consider only one spatial quantum number n. Three-dimensional problems involve three spatial quantum numbers, one for each dimension.

Spin one-half particles such as electrons are called **fermions**; they obey the exclusion principle. Other particles with either zero or integral spin quantum numbers, such as α particles, deuterons, and photons, are called **bosons**; these do *not* obey the exclusion principle.

Most systems we will consider involve a large number N of electrons. Let us look at these N electrons in a one-dimensional box at $T = 0$. Only two of the N electrons can be in the lowest energy state corresponding to $n = 1$. There will then be two electrons in the next higher state $n = 2$, two electrons in the state $n = 3$, and so on, until all the electrons are accounted for, and all the available energy states are filled. For a very large number of electrons, the electrons in the highest state can have an energy much larger than that of the lowest ground-state energy E_1.

For N electrons, the corresponding value of n of the highest electron state is $n = N/2$. This corresponds to the **Fermi energy** E_F given by setting $n = N/2$ in the above expression for quantized energy levels in a one-dimensional box: $E_F = \left[h^2/(32m_e)\right](N/L)^2$. For three dimensions, the Fermi energy at $T = 0$ is given by $E_F = \left[h^2/(8m_e)\right]\left[3N/(\pi V)\right]^{2/3}$, where V is the volume of the conductor. When numerical values are substituted, the expression for E_F becomes $E_F = (0.3646 \text{ eV} \cdot \text{nm}^2)(N/V)^{2/3}$.

The above expressions show that the Fermi energy at $T = 0$ depends upon N/V, the number of free electrons per unit volume, called the **number density of free electrons**, which is a characteristic of a given material. In turn, for a given material there is a given Fermi energy E_F at $T = 0$, which can be looked up in standard tables.

The Fermi energy is the highest energy of the N electrons at $T = 0$ in the three-dimensional conductor under consideration. The N electrons in the conductor of volume V have energies ranging from the ground-state energy E_1 to the Fermi energy E_F. The average energy of this system can be shown to be $E_{av} = \frac{3}{5}E_F$. For typical conductors, E_{av}, which is of the order of several eV, is much larger than thermal energies of about $kT = 0.026$ eV at a room temperature of $T = 300$ K.

The **Fermi factor** $f(E)$ gives the probability of finding an electron in a given electron state. At $T = 0$, all the states below E_F are filled, and all the states above E_F are empty. Therefore, the probability is equal to 1 for finding an electron in a state below E_F and is equal to zero for finding an electron in a state above E_F. as shown.

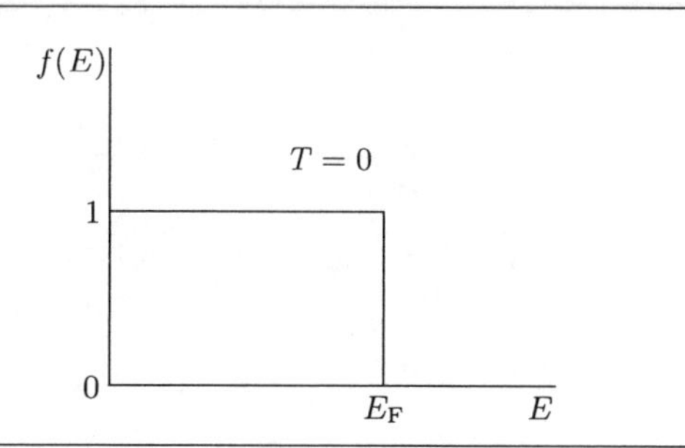

At temperatures greater than $T = 0$ the picture changes, but not by much. Electrons with energies around the Fermi energy E_F gain extra energy of the order of kT, which at 300 K is only 0.026 eV. Thus there will be a certain probability of finding these electrons with an energy slightly higher than the Fermi energy at $T = 0$, and the Fermi factor for $T > 0$ looks as shown. Since for $T > 0$ there is no distinct energy separating filled and unfilled states, the Fermi energy at $T > 0$ is defined to be that energy for which the probability of being occupied is one-half. Except for very high temperatures, the Fermi energy for $T > 0$ is very close to the Fermi energy at $T = 0$. The **Fermi temperature** T_F is defined by $kT_F = E_F$. Note that even at $T = 0$ there is a finite Fermi temperature. For most metals, the Fermi temperature is so large that the metal would not be a solid at actual temperatures around T_F.

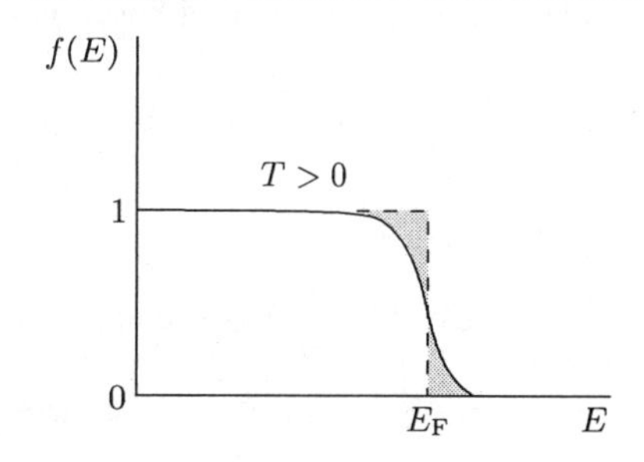

Related to the Fermi energy is the **Fermi speed** $u_F = \sqrt{2E_F/m_e}$.

When two different metals are placed in contact with each other, a potential difference $V_{contact}$ called the **contact potential** develops between them. The contact potential depends upon the work function ϕ of each of the metals. Recall from the photoelectric effects discussed in Chapter 34 that ϕ is the energy required to remove the least tightly bound electrons from a metal. The contact potential is related to the work functions ϕ_1 and ϕ_2 of the two metals in contact by $V_{contact} = (\phi_1 - \phi_2)/e$.

Important Derived Results

Quantized energies of a particle in a one-dimensional box	$E_n = n^2 \frac{h^2}{8mL^2} = n^2 E_1$, for $n = 1, 2, 3 \ldots$
Fermi energy at $T = 0$ in one dimension	$E_F = \frac{h^2}{32m_e}\left(\frac{N}{L}\right)^2$
Fermi energy at $T = 0$ in three dimensions	$E_F = \frac{h^2}{8m_e}\left(\frac{3N}{\pi V}\right)^{2/3}$
Fermi energy at $T = 0$ in three dimensions	$E_F = (0.3646\ \text{eV} \cdot \text{nm}^2)(N/V)^{2/3}$
Average energy of electrons in a Fermi gas at $T = 0$	$E_{av} = \frac{3}{5}E_F$
Fermi temperature	$kT_F = E_F$
Relation between Fermi speed and Fermi energy	$u_F = \sqrt{\frac{2E_F}{m_e}}$
Relation of contact potential to work function	$V_{contact} = (\phi_1 - \phi_2)/e$

Common Pitfalls

> Understand the difference between the general Pauli exclusion principle and the Pauli exclusion principle for spatial states.

> Understand the difference between fermions and bosons.

5. TRUE or FALSE: In the ground state, N bosons in a one-dimensional box have the same energy as N fermions in the same box.

6. Why does the classical model fail in explaining the observed results of electrical conduction?

e⁻ have wave characteristics, obey Pauli, & follow Fermi-Dirac.

Try It Yourself #3

The density of potassium is 0.870 g/cm^3 and its atomic mass is 38.96. Assuming that each potassium atom contributes one valence electron, calculate the Fermi energy for potassium.

Picture: Avogadro's number can be used to find the number density of potassium atoms, and hence electrons. Then use the expression relating the Fermi energy to the number density of electrons.

Solve:

Find the number of potassium atoms per unit volume.	
If each potassium atom provides one free electron, the number density of free electrons is equal to the number density of atoms.	
Relate the Fermi energy to the number density of electrons.	$E_F = 2.07$ eV

Check: Given sample values in the textbook, we should expect a Fermi energy in the range of a few eV, which we found.

38.4 Quantum Theory of Electrical Conduction

In a Nutshell

If we use the Fermi speed in place of the average speed, the expression for the resistivity of a metal becomes $\rho = m_e u_F/(n_e e^2 \lambda)$. Unfortunately, the Fermi speed is considerably larger than the average speed, which means the calculated resistivity above is now about 100 times greater than the experimentally determined value. To solve both this problem and the temperature dependence problem of the resistivity, we need to re-interpret the mean-free-path.

Quantum-mechanically, electron waves do not scatter at all in a perfectly ordered crystal. It is only *imperfections* in the crystal ordering that cause scattering to occur. As a result, we can still use $\lambda = 1/(n_{ion}A) = 1/(n_{ion}\pi r_0^2)$ for the mean free path, but only if we interpret r_0 to be the amplitude of thermal vibrations, rather than the radius of the ions in the lattice.

Important Derived Results

Quantum-mechanical mean free path $$\lambda = \frac{1}{n_{ion}A} = \frac{1}{n_{ion}\pi r_0^2}$$

38.5 Band Theory of Solids

In a Nutshell

The free-electron picture of a Fermi gas of electrons interacting with atoms in the lattice does not explain why some materials (conductors) are good conductors of electricity, why other materials (insulators) are poor conductors of electricity, and why still other materials (semiconductors) have conductive properties somewhere between those of conductors and insulators. Also, the free-electron picture does not explain why the ratio of resistivities of insulators and conductors can be much larger than 10^{23}.

An explanation of these and other results is given by the **band theory of solids**. The electrons in a single atom in its ground state occupy the lowest-lying energy levels, with higher energy levels being unoccupied. Figure (a) shows the first three energy levels of an atom (designated as 1s, 2s, and 2p). If two of these atoms are brought close together, the interaction between them shifts the energy of each level for each atom; the result is that each of the three previously single energy levels is split into two slightly different levels, as shown in Figure (b). When N atoms are brought close together, their interactions cause each previously single energy level to split into N separate but closely spaced levels. Figure (c) shows the splittings when six atoms are brought together. Now consider a macroscopic solid, in which a large number of atoms are very close together and N is of the order 10^{23}. The N energy levels are now so close together that they can be regarded as forming continuous bands, one band for each of the previously single energy levels, as shown in Figure (d).

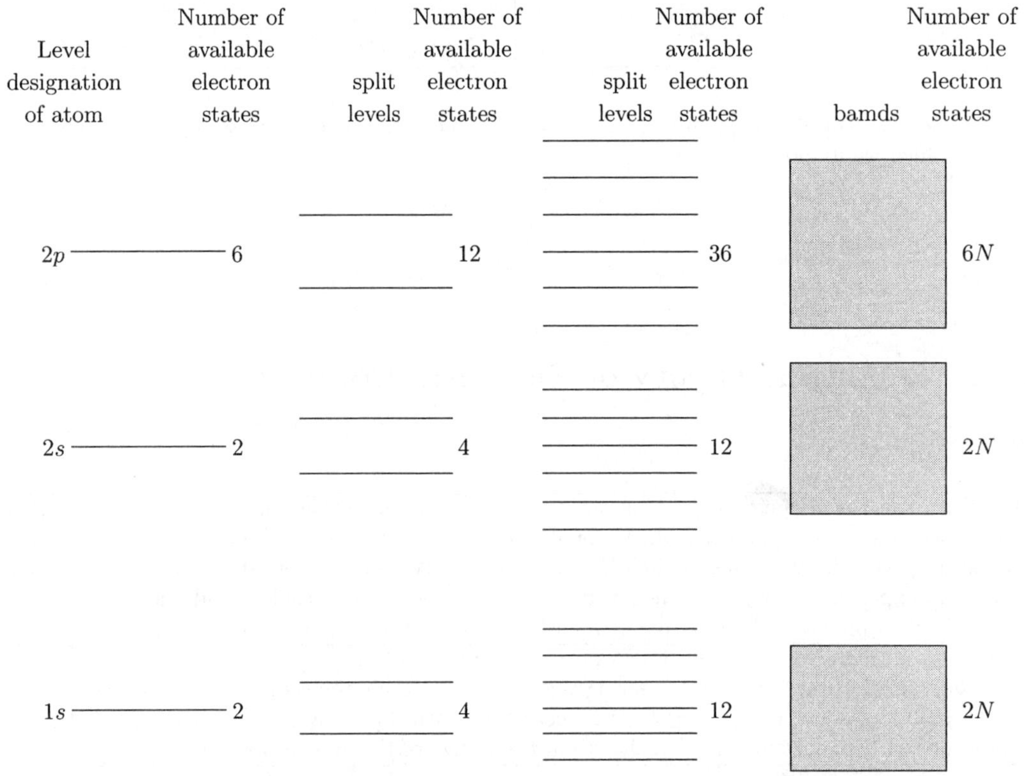

Level designation of atom	Number of available electron states	split levels	Number of available electron states	split levels	Number of available electron states	bamds	Number of available electron states
$2p$	6		12		36		$6N$
$2s$	2		4		12		$2N$
$1s$	2		4		12		$2N$

(a) One isolated atom (b) Two interacting atoms (c) Six interacting atoms (d) N interacting atoms

The positions of these energy-level bands relative to each other and how they fill with electrons determine whether a material is a conductor, insulator, or semiconductor. A band can be filled with electrons, partially filled with electrons, or empty, as shown below. The band occupied by the highest-energy valence electrons is called the **valence band**. The lowest band in which there are unoccupied energy states is called the **conduction band**. An energy gap between allowed bands is called a **forbidden energy band** or **energy gap**.

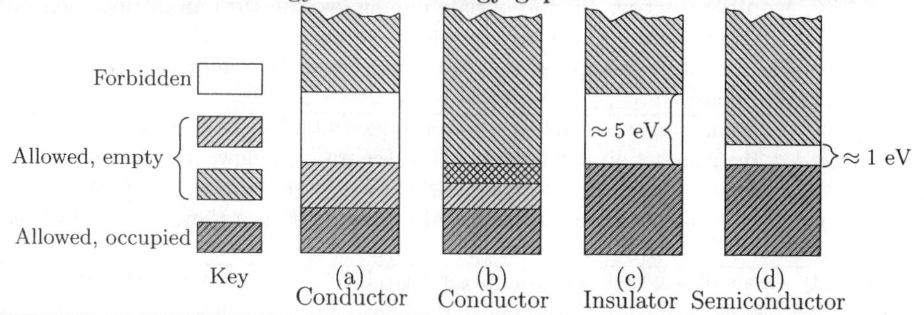

A **conductor** can result in two ways. If the valence band is only partially filled, as in Figure (a) above, it is easy to excite valence electrons into the energy levels in the unfilled part of the band. We simply apply an electric field to give these electrons the kinetic energy they need to participate in electrical conduction. In this case, the valence band is also the conduction band. In another situation, shown in Figure (b), the conduction band overlaps the partially or totally filled valence band, resulting in unoccupied energy levels into which valence electrons can move to participate in electrical conduction.

The valence band in an **insulator** is completely filled with electrons, and the conduction band is separated from the valence band by a large energy gap, of the order of 5 eV, as in Figure (c). At ordinary temperatures around 300 K, the thermal energy is of the order of 0.026 eV, so only a few electrons have enough energy to be in the conduction band. Even if we apply an electric field to an insulator, we cannot accelerate the electrons to higher kinetic energies because there are no nearby energy levels for them to occupy. Thus little electrical conduction results.

In a **semiconductor** the energy gap between a filled valence band and the conduction band is small, of the order of 1 eV, as in Figure (d). At ordinary temperatures, thermal excitation has promoted a sizable number of electrons into the conduction band, and each of these thermally excited electrons has left behind it an unfilled **hole** in the valence band. This type of material is called an **intrinsic semiconductor**. When we apply an electric field, the thermally excited electrons acquire kinetic energy and participate in electrical conduction because many energy levels are available to them; electrons in the valence band also can acquire kinetic energy and participate in electrical conduction because there are holes in the energy levels into which they can move. The movement of electrons into holes is equivalent to positive charges moving in a direction opposite to that of the electrons.

For conductors, resistivity increases with increasing temperature and conductivity (the reciprocal of resistivity) decreases. But for semiconductors, resistivity decreases with increasing temperature and conductivity increases. When we raise the temperature of a semiconductor, we excite more electrons into the conduction band, leaving more holes in the valence band. Because both electrons and holes can participate in electrical conduction, conductivity increases and resistivity decreases.

38.6 Semiconductors

In a Nutshell

The properties of intrinsic semiconductors can be changed in a controllable way by introducing impurities into the semiconductor—that is, by **doping** the intrinsic semiconductor to turn it into an **impurity semiconductor**. By appropriate doping, we can turn an intrinsic semiconductor into either an ***n*-type semiconductor**, in which conduction is due primarily to negatively charged electrons, or into a ***p*-type semiconductor**, in which conduction is due primarily to positively charged holes.

The impurity atoms of an n-type semiconductor produce additional electrons that occupy energy levels just below the conduction band, as shown in Figure (a) below. These electrons can easily move into the conduction band to participate in electrical conduction, without leaving extra holes in the valence band. Such levels are called **donor levels** because they donate electrons to the conduction band.

The impurity atoms of a p-type semiconductor introduce a deficiency of electrons and produce unoccupied energy levels, or holes, just above the nearly filled valence band, as shown in Figure (b) below. Electrons in the valence band can easily move into these holes, thereby producing extra holes in the valence band to participate in electrical conduction, without producing extra electrons in the conduction band. Such levels are called **acceptor levels** because they accept electrons from the valence band.

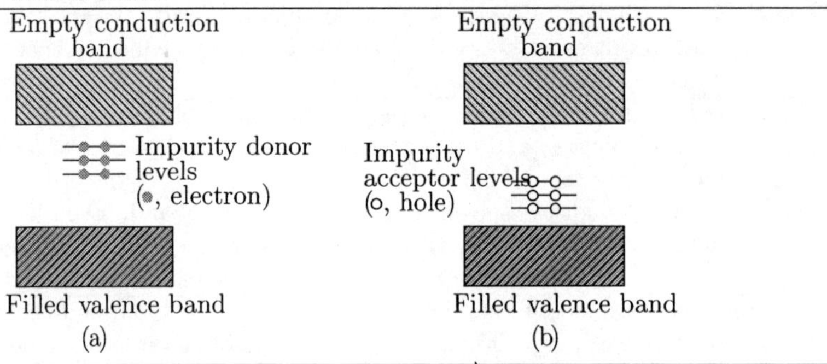

Common Pitfalls

> Understand the band structure that leads to the differences between conductors, insulators, and semiconductors.

> Remember the difference between n-type and p-type semiconductors.

7. TRUE or FALSE: The band theory arises because atoms in metals attract each other as though rubber bands were stretched between them.

8. Explain the difference between an insulator and a semiconductor.

Insulator → Valence band completely full.

Semiconductor → E gap b/t valence & conduction band is small.

38.7 Semiconductor Junctions and Devices*

In a Nutshell

A p-type semiconductor has a larger concentration of positively charged holes that are available for electrical conduction than of available negatively charged electrons; an n-type semiconductor has a larger concentration of available electrons. A ***pn* junction** is formed when a p-type and an n-type semiconductor are placed in contact. Because of the unequal concentrations of electrons and holes, electrons diffuse from the n to the p side and holes diffuse from the p to the n side, until the attractive forces between the holes and electrons result in an equilibrium configuration. (Both the holes and the electrons are called charge carriers.)

At equilibrium there is a double layer of charges at the junction: negative charges on the p side and positive charges on the n side as shown. Between the negative and positive layers is the depletion region, which has no charge carriers. As a consequence there is a potential difference V across the junction, with the n side being at the higher voltage.

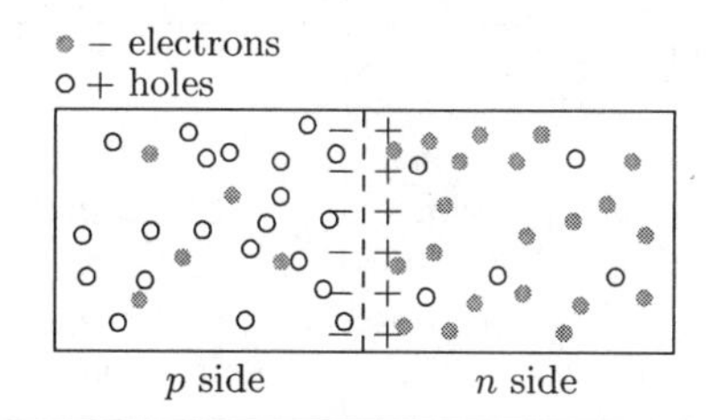

Examine what happens when we connect an external voltage source across the junction. If we connect the positive terminal of the source to the p side of the junction, we obtain a **forward bias**, as shown in Figure (a)—that is, the external potential difference reduces the potential difference across the junction, producing an increased diffusion of electrons and holes; this increased diffusion results in a current that increases exponentially with the applied external voltage. If we connect the positive terminal of the source to the n side of the junction, we obtain a **reverse bias**, as in Figure (b). In this case the external potential difference adds to the original junction potential difference V, producing a very small reverse current that eventually reaches a saturation value as the external voltage is increased. The junction thus acts as a **diode**, through which conduction takes place essentially in only one direction. Other types of devices built on pn junctions are **tunnel diodes**, **solar cells**, **light-emitting diodes** (LEDs), and **semiconductor lasers**.

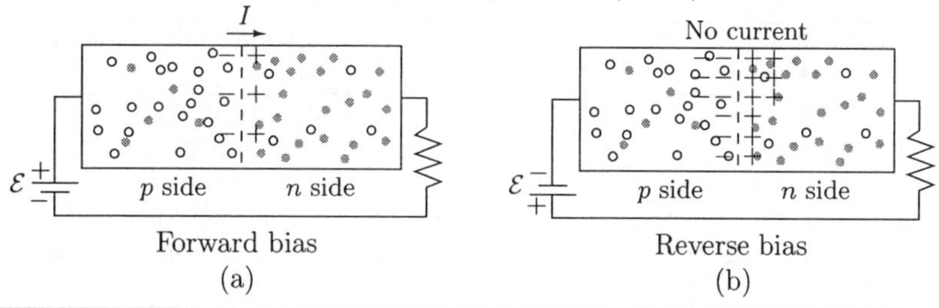

Forward bias
(a)

Reverse bias
(b)

A **junction transistor** consists of a very thin semiconductor region sandwiched between two semiconductor regions of the opposite type. A *pnp* transistor and an *npn* transistor are shown below with their circuit symbols. The narrow central region of a transistor is the **base**, and the outer regions are the **emitter** and **collector**. Thus a transistor consists of two pn junctions, one between emitter and base and the other between base and collector. The emitter is much more heavily doped than the base and collector. We will discuss only the operation of a *pnp* transistor; the operation of an *npn* transistor is essentially the same.

*Optional material

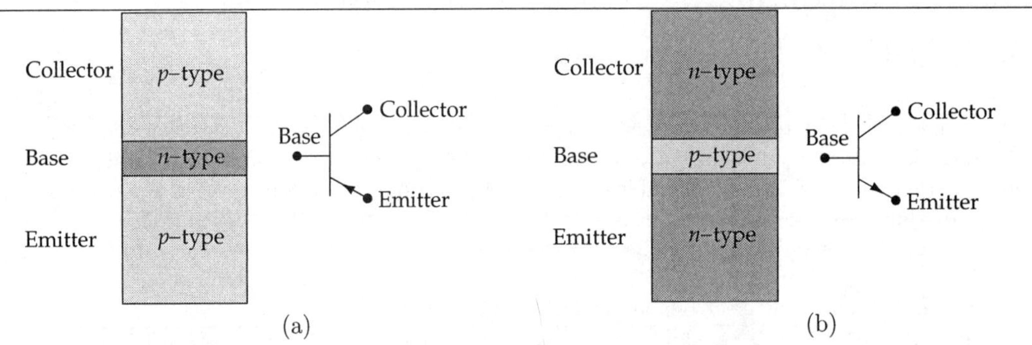

The basic operation of a transistor can be seen from an analysis of the figure at right. The voltage V_{ec} causes forward biasing of the emitter-base junction and reverse biasing of the base-collector junction. Because of the forward bias, the heavily doped p-type emitter emits holes that flow across the emitter-base junction into the base, resulting in the emitter current I_e. A small number of the holes that flow across the emitter-base junction will tend to recombine in the base, but this would result in an undesirable accumulation of positive charge in the base that would prevent other holes from crossing the junction. To prevent this accumulation, an alternate path is provided to draw off these holes by connecting a battery V_{eb} between the base and emitter and letting a small base current I_b pass. The collector current is then $I_c = I_e - I_b$.

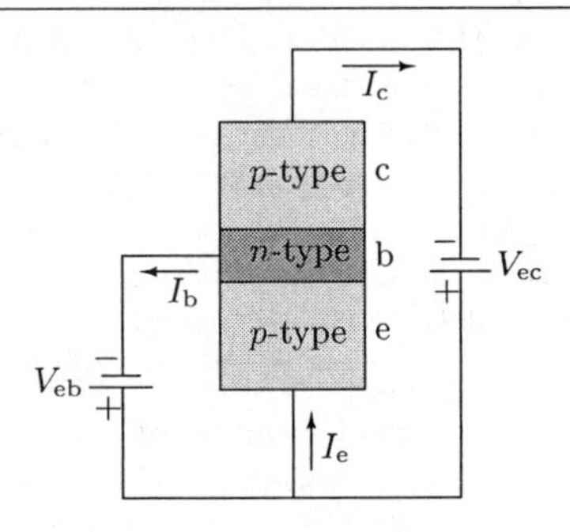

In practice I_b is much smaller than either I_c or I_e. The **current gain** β of the transistor is defined by $I_c = \beta I_b$. Typically β is in the range from 10 to 100.

A common use for a transistor is the amplification of small time-varying signals. The operation of a typical amplifier circuit can be seen from an analysis of the figure below. The small time-varying input signal v_s produces a time-varying current i_b that is added to the steady-state current I_b. This results in a large time-varying output current $i_c = \beta i_b$ that is added to the steady-state current I_c in the collector. The current i_c produces a time-varying output voltage v_L across the load resistance R_L.

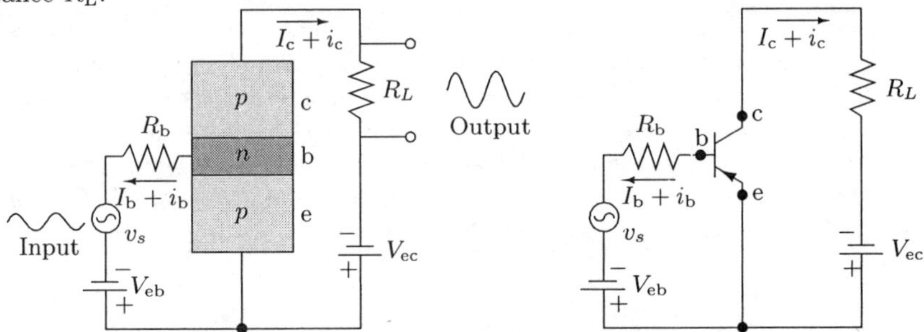

The ratio of the output voltage v_L to the input voltage v_s is the **voltage gain** of the amplifier. If r_b is the internal resistance of the transistor between the base and emitter, and R_b is the resistance in series with the bias voltage V_{eb}, we have Voltage gain $= v_L/v_s = \beta R_L/(R_b + r_b)$.

An **integrated circuit** (IC) chip is a collection of resistors, capacitors, diodes, and transistors, interconnected through circuits, that is fabricated on a single semiconductor crystal, usually silicon. It is possible to produce ICs that contain several hundred thousand components in an area smaller than 1 cm^2. The development of ICs since the early 1960s has revolutionized the electronics industry and has led to thousands of applications in computers, cameras, watches, communication networks, and many other areas.

Important Derived Results

Current gain of a transistor	$I_c = \beta I_b$
Voltage gain of a transistor	$\text{Voltage gain} = \dfrac{v_L}{v_s} = \beta \dfrac{R_L}{R_b + r_b}$

Common Pitfalls

> Remember the difference between n-type and p-type semiconductors.

9. TRUE or FALSE: A transistor is equivalent to a pn junction joined to an np junction.

10. What is the difference between the forward and the reverse biasing of a pn-junction diode?

Forward bias → (+) to p = large current

Reverse bias → (+) to n ≈ no current

38.8 Superconductivity

In a Nutshell

The resistivity of a **superconductor** drops suddenly to zero when its temperature is decreased below a **critical temperature** T_c, as shown, which varies from one superconductor to another. Below T_c the resistivity is truly zero. If a current is established in a superconducting loop, it persists for years with no measurable decay. When superconductivity was first discovered in the early 1900s, materials became superconducting only at very low temperatures of the order 0.1 to 10 K. At present, though, many **high-temperature superconductors** have been found with critical temperatures as high as 138 K at atmospheric pressure. This has revolutionized the study and applications of superconductors because high-temperature superconductors can be cooled with inexpensive liquid nitrogen, which boils at 77 K.

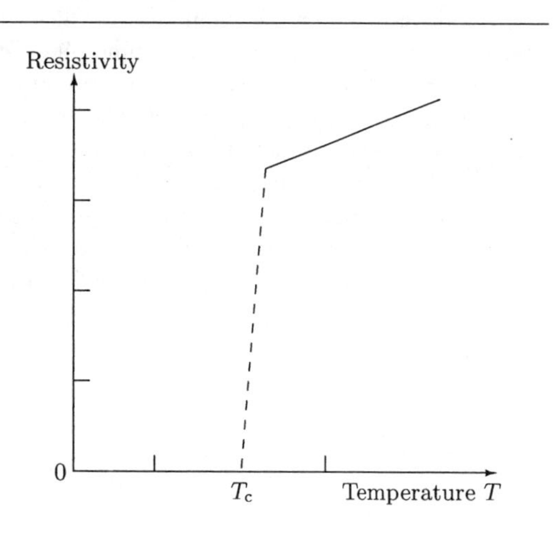

The **BCS theory** is named after John Bardeen, Leon Cooper, and Robert Schrieffer, who first put forth an explanation of superconductivity in 1957, and who received the Nobel Prize for their work in 1972. In the BCS theory, two electrons in a superconductor at low temperature form a bound state, called a **Cooper pair**, as a result of interactions of the electrons with the crystal lattice of the material. This can be seen in an intuitive manner as follows. The figure shows positive ions in the lattice being attracted toward an electron. The effect of this interaction is to produce a net positive charge in the region of these ions, which can now attract another electron toward the region. Thus the lattice acts as a mediator that produces an attractive force between the two electrons in a Cooper pair that exceeds their Coulomb repulsion.

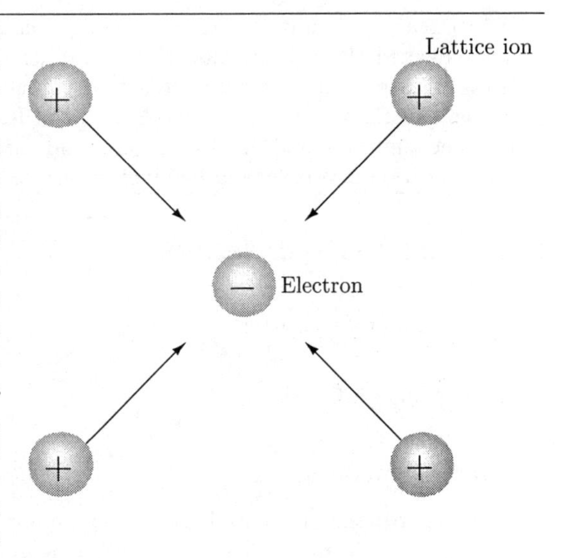

Electrons in a Cooper pair have opposite spins and have equal and opposite linear momenta. Thus a Cooper pair has the properties of a single particle with zero momentum and zero spin. This means that a Cooper pair behaves like a boson, which does not obey the Pauli exclusion principle, so there can be any number of Cooper pairs in the same quantum state at the same energy.

When a material is in a superconducting state, its Cooper pairs can absorb (or emit) energy only by breaking up (or forming). (This is analogous to molecules that must absorb energy before they can break up into their constituent atoms.) The energy required to break up a Cooper pair is called the **superconducting energy gap** E_g, which is of the order of 10^{-3} eV. In the BCS theory, E_g at $T = 0$ is related to the critical temperature T_c by $E_g = (7/2)kT_c$. As the temperature is increased from $T = 0$, E_g decreases, reaching zero at $T = T_c$, when superconductivity ceases.

When electrical conduction occurs in a superconductor, all the Cooper pairs have the same momentum, which remains constant because individual pairs cannot be scattered by the lattice ions. Because the lattice ions cannot scatter Cooper pairs, there is no resistance.

Consider a junction consisting of two superconducting materials separated by a layer of insulating material that is only a few nanometers thick. This is called a **Josephson junction**. Even when no voltage is applied across this junction, Cooper pairs tunnel through it. The resulting dc current, called the **dc Josephson effect current**, is given by $I = I_{max}\sin(\phi_2 - \phi_1)$, where the maximum current I_{max} depends on the thickness of the junction. The angles ϕ_1 and ϕ_2 are the phases of the wave functions of the Cooper pairs in the two superconductors.

If a dc voltage V is applied across the junction, the result is, somewhat surprisingly, an ac current called the **ac Josephson effect current**, with frequency $f = 2eV/h$. Because we can measure frequencies extremely accurately, this effect provides an experimental method for measuring the ratio e/h very precisely, and is also used to establish precise voltage standards.

Important Derived Results

Superconducting energy gap at $T = 0$	$E_g = \frac{7}{2}kT_c$
dc Josephson effect current	$I = I_{max}\sin(\phi_2 - \phi_1)$
ac Josephson effect current frequency	$f = \frac{2e}{h}V$

Common Pitfalls

> Do not confuse Cooper pairs in the BCS theory with free electrons in the free-electron theory.
> Understand the differences between the dc and ac Josephson effects.

11. TRUE or FALSE: A Cooper pair has the properties of a spin-$\frac{1}{2}$ fermion. → spin-0 boson

12. What is the critical temperature of a superconductor?

$E_g = (7/2)kT_c \rightarrow T_c = \frac{E_g}{(7/2)k}$

Try It Yourself #4

The critical superconducting temperature for lead is 7.19 K. Find the energy in eV required to break up a lead Cooper pair at $T = 0$.

Picture: The energy required to break up the pair is equal to the superconducting energy gap.

Solve:

Find the superconducting energy gap.	$E_g = 2.17 \times 10^{-3}$ eV

Check: From reading the text, we should expect values of E_g to be significantly less than 1 eV.

38.9 The Fermi–Dirac Distribution

In a Nutshell

The **Fermi–Dirac distribution function** $n(E)$ gives the number of electrons dN having energies in the interval between E and $E+dE$ according to $dN = n(E)\,dE$. The expression for $n(E)$ is composed of two parts, the density of states and the Fermi factor.

The number of states between E and $E+dE$ is $g(E)\,dE$, where $g(E)$ is the **density of states** given by

$$g(E) = \frac{8\pi\sqrt{2}m_e^{3/2}V}{h^3}E^{1/2} = \frac{3N}{2E_F^{3/2}}E^{1/2}$$

The density of states does not depend on temperature. A plot of $g(E)$ versus E is shown.

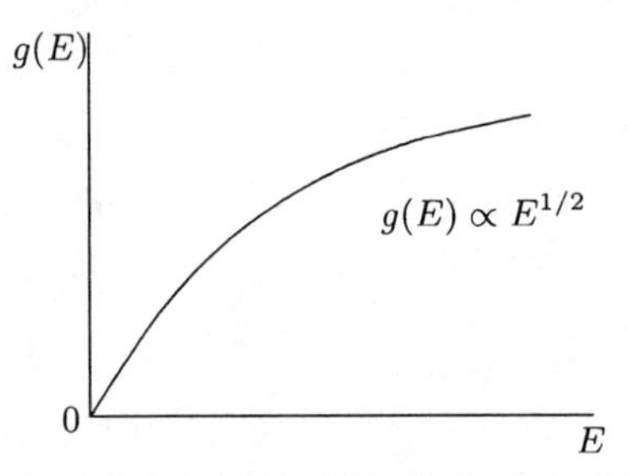

The other part of the overall distribution comes from the Fermi factor $f(E)$, which gives the probability of finding an electron in a given state. We have seen in Section 38-3 that at $T = 0$ the Fermi factor is

$$f(E) = \begin{cases} 1 & E \le E_F \\ 0 & E > E_F \end{cases}$$

For $T > 0$ the Fermi factor takes on a more complicated form given by

$$f(E) = \frac{1}{e^{(E-E_F)/(kT)}+1}$$

Plots of the Fermi factor at $T = 0$ compared with the Fermi factor at $T > 0$ are shown on page 85.

The number of electrons in a given energy interval dE is the number of states in the energy interval, $g(E)\,dE$, multiplied by the probability of finding an electron in a given state, $f(E)$: $n(E)\,dE = g(E)\,dE\,f(E)$. Using the expressions for $g(E)$ and $f(E)$ we obtain $n(E)$, which is the Fermi–Dirac distribution function:

$$n(E) = \frac{8\pi\sqrt{2}m_e^{3/2}V}{h^3}\frac{E^{1/2}}{e^{(E-E_F)/(kT)}+1}$$

Here we show a plot of $n(E)$. The dashed curve shows $n(E)$ at $T = 0$, where there are no electrons with energies larger than the Fermi energy. At higher temperatures, some electrons with energies near the Fermi energy are excited above the Fermi energy, as indicated by the shaded region. Because only those electrons within about kT of the Fermi energy can be excited to higher energy states, the difference between $n(E)$ at temperature T and at $T = 0$ is very small except for extremely high temperatures.

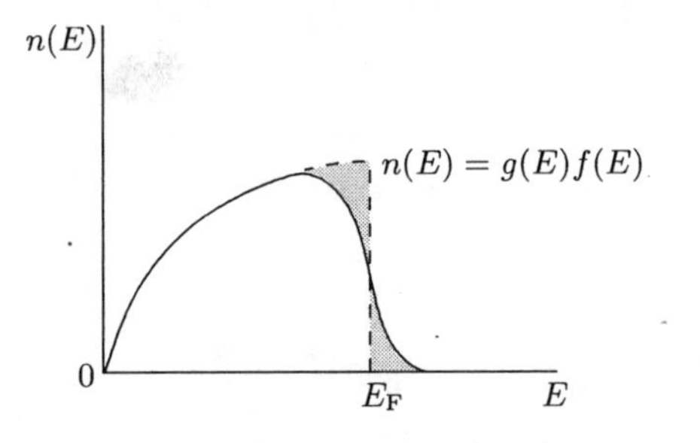

Important Derived Results

Fermi factor at $T = 0$	$f(E) = \begin{cases} 1 & E \le E_F \\ 0 & E > E_F \end{cases}$
Fermi factor at $T > 0$	$f(E) = \dfrac{1}{e^{(E-E_F)/(kT)} + 1}$
Number of electrons with energies between E and $E + dE$	$dN = n(E)\, dE$
Density of states	$g(E) = \dfrac{8\pi\sqrt{2}m_e^{3/2}V}{h^3}E^{1/2} = \dfrac{3N}{2E_F^{3/2}}E^{1/2}$
Fermi–Dirac distribution function	$n(E) = \dfrac{8\pi\sqrt{2}m_e^{3/2}V}{h^3}\dfrac{E^{1/2}}{e^{(E-E_F)/(kT)} + 1}$

Common Pitfalls

- It can be easy to confuse the number of particles that have an energy between E and $E + dE$ with the number of states that have an energy between E and $E + dE$.
- It can be easy to confuse the number of particles that have an energy between E and $E + dE$ with the Fermi–Dirac distribution function $n(E)$.

13. TRUE or FALSE: The density of states $g(E)$ is the same at a higher temperature as at $T = 0$.

14. How does the Fermi–Dirac distribution function vary with temperature?

Inversely w/ T $\Longrightarrow f(E) = \dfrac{1}{e^{(E-E_F)/(kT)} + 1}$

Try It Yourself #5

In a conductor of volume V containing N free electrons, the density of states is $g(E) = 3NE^{1/2}/(2E_F^{3/2})$, where E is measured from the bottom of the conduction band. Find the number of states from the bottom of the conduction band to the Fermi energy.

Picture: Find the number of states in a differential energy interval dE. Integrate this expression from zero to the Fermi energy.

Solve:

Determine the number of states in a small energy interval.	

Integrate the expression from step one to find the total number of states.	$N_S = N$

Check: Since electrons obey the Pauli exclusion principle, the N free electrons in the conduction band must occupy N states, which is exactly what we found.

Try It Yourself #6

Find the ratio of the number of free electrons in silver that are in a small energy interval dE located at 0.1 eV above and 0.1 eV below the Fermi level of 5.50 eV at a temperature of 300 K.

Picture: This will be the ratio of the Fermi–Dirac distribution function at each of the two energies.

Solve:

Write an *algebraic* expression for the ratio of the two Fermi-Dirac distribution functions.	
Evaluate each of the exponents.	
Now the ratio can be calculated.	$\frac{n(E_U)}{n(E_L)} = 0.0212$

Check: You should expect relatively fewer electrons at the higher energy level, which is what we found.

QUIZ

1. TRUE or FALSE: The number of electrons occupying energy levels between E and $E + dE$ is equal to the number of states in the same energy interval.

2. TRUE or FALSE: As a general rule, it is advantageous to get rid of impurities in a semiconductor.

3. Explain the difference between a conductor and an insulator.

4. What is the Fermi factor?

5. Describe the differences between the dc and ac Josephson effects.

6. How many free electrons are below the Fermi energy in 5.00 cm^3 of copper at $T = 0$? The number density of electrons in copper is 8.47×10^{22} electrons/cm^3 and the Fermi energy is 7.04 eV.

7. The ratio of the number of free electrons in a material that are in a small energy interval dE located 0.100 eV above and 0.100 eV below the Fermi level of the material is 0.02113 at $T = 300$ K. Find the Fermi level of the material, and from this value identify the material.

Chapter 39

Relativity

39.1 Newtonian Relativity

In a Nutshell

To measure distances and times an observer needs a set of coordinate axes to measure distance and a set of synchronized clocks to measure time. The coordinate axes and clocks make up a **reference frame**.

Reference frames are used to specify the location of events in space and time. An **event** might be the striking of a lightning bolt, the collision of two particles, or the explosion of a distant supernova. To specify an event you assign it four space-time coordinates (x, y, z, t). The three position coordinates x, y, and z determine the distance from the origin of your reference frame to the event. The time coordinate t, measured on a clock stationary in your reference frame that is at the location of the event, tells the instant when the event takes place.

An **inertial reference frame** is a reference frame in which objects move with a constant velocity when no net force acts on them, in accordance with Newton's laws. A spaceship coasting in outer space far away from any planets or stars is an inertial reference frame. Release an object in the spaceship and it will remain motionless in front of you. If you want to accelerate the object, you have to exert forces on it.

If a spaceship's rockets are fired, causing it to accelerate, the spaceship is no longer an inertial reference frame; it is an **accelerated reference frame**. Release an object in an accelerating spaceship and the object will appear to accelerate toward or away from you without a net force acting on it. A rotating merry-go-round is also an accelerated reference frame. Place a marble on the floor of the merry-go-round and it will accelerate toward the perimeter, even though the net force acting on the marble appears to be zero.

According to Newtonian (and Einsteinian) relativity, a reference frame moving at a constant velocity relative to a known inertial reference frame is also an inertial reference frame. A consequence of this statement is the principle of Newtonian relativity, "Absolute motion cannot be detected."

Before Einstein put forth his theory of special relativity in 1905 it was generally believed that light and other electromagnetic phenomena did not obey a principle of relativity. According to Newtonian concepts the speed of light $c = 3.00 \times 10^8$ m/s can be measured only by a "privileged observer" who is at rest in "absolute space" in which there is a medium called the **ether** that supports the propagation of light and other electromagnetic waves. Other observers moving relative to this privileged observer will necessarily measure the speed of light to be different from the value c, and consequently will also find Maxwell's equations to have a form different from what the privileged observer obtains. Michelson in 1881 and later Michelson and Morley in 1887 developed an interferometer sensitive enough to measure the expected change of the speed of light as Earth moved in different directions through the hypothesized ether. The result of these and many other subsequent experiments was that the speed of light did not change, and no motion of Earth through the ether was ever detected. This is one of the most famous null experimental results ever obtained.

Common Pitfalls

1. TRUE or FALSE: The medium in which light was (mistakenly) assumed to propagate was called the ether.

2. Was the Michelson and Morley experiment devised primarily to make very precise measurements of the speed of light? Explain.

39.2 Einstein's Postulates

In a Nutshell

Einstein's theory of special relativity follows from two postulates that unite all realms of physics—mechanics, electromagnetism, nuclear physics, and every other field of physics—under the common umbrella of relativity.

Postulate 1: "Absolute uniform motion cannot be detected." This is simply an extension of Newtonian relativity to include all types of physical measurements, not just those that are mechanical in origin.

Postulate 2: "The speed of light is independent of the motion of the source," or alternatively, "Every observer measures the same value c for the speed of light." This second postulate expresses a property of waves that you are probably familiar with from your previous studies. When a wave such as a sound wave leaves its source, its movement through the propagating medium is independent of how the source is moving. In particular, the speed with which the wave moves through the medium is not related to how fast or slow the source is moving.

The alternate statement of Einstein's second postulate requires a bit of contemplation, as it runs counter to commonsense (Newtonian) concepts of how velocities add. One thing that follows from this alternate form of the second postulate is an explanation of the null result of the Michelson-Morley experiment. If all observers measure the same value for the speed of light, it follows that no experiment will ever show two observers measuring different values for the speed of light.

39.3 The Lorentz Transformation

In a Nutshell

An important question in both Newtonian and Einsteinian relativity is how the coordinates assigned to a particular event by two observers in different inertial frames moving relative to each other are interrelated. The figure shows a reference frame S' moving with a velocity $\vec{v}'$ relative to a reference frame S along the collinear x-x' axes. Along with the rectangular coordinate axes xyz in reference frame S is a set of stationary synchronized clocks that measure the time t. A different set of synchronized clocks measuring the time t' is located at rest relative to the rectangular coordinate axes $x'y'z'$ of reference frame S'.

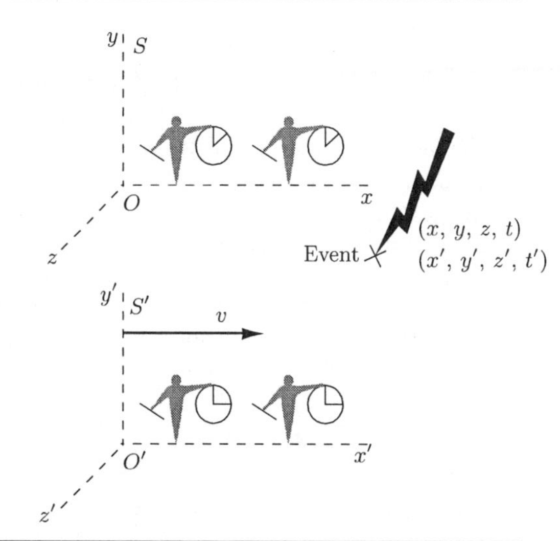

Suppose an observer in each reference frame observes a particular event, say, the striking of a lightning bolt at a given instant. Each observer describes where and when the event takes place by assigning it four space-time coordinates: (x, y, z, t) are assigned in S, and (x', y', z', t') are assigned in S'. The interrelationship between these sets of coordinates is called a **coordinate transformation.**

Using his postulates, Einstein found that the space-time coordinates measured in S and S' are related by the **Lorentz transformation** given at right. In these expressions $\gamma = 1/\sqrt{1 - v^2/c^2}$ and v is the velocity of S' along the common x-x' axis relative to S. v is positive if S' is moving in the positive x-x' direction and negative if S' is moving in the negative x-x' direction.

$$x = \gamma(x' + vt') \qquad x' = \gamma(x - vt)$$
$$y = y' \qquad y' = y$$
$$z = z' \qquad z' = z$$
$$t = \gamma(t' + vx'/c^2) \qquad t' = \gamma(t - vx/c^2)$$

If v is much smaller than the speed of light ($c = 3.00 \times 10^8$ m/s), then $\gamma \to 1$ and the Lorentz transformation reduces to the classical **Galilean transformation.**

$$x = x' + vt' \qquad x' = x - vt$$
$$y = y' \qquad y' = y$$
$$z = z' \qquad z' = z$$
$$t = t' \qquad t' = t$$

In general, two events will take place at different locations. Two separated clocks are then needed in a reference frame to measure the time interval between the two events, one clock located at one event and a second clock located at the other event. When two events occur at the *same* place in a particular reference frame, the time interval between the two events can be measured with a single clock located at the position where the two events occur. This time interval Δt_{p} measured by a single clock is called the **proper time** between the two events. All other observers moving relative to the reference frame in which proper time is measured will find the time interval Δt between the two events, as measured by their clocks that are necessarily separated, to be larger than Δt_{p} according to the **time dilation expression** $\Delta t = \Delta t_{\mathrm{p}}/\sqrt{1 - v^2/c^2}$.

An unaccelerated object can be at rest in only one inertial reference frame. An observer at rest in this particular inertial frame can measure the spatial coordinates of the end points of the object at any time. The length of the object measured in the reference frame in which the object is at rest is called its **proper length** L_{p}. Consider a rod moving along the x-x' axis with a velocity v relative to frame S. An observer in S who wishes to determine the length L of the moving rod must measure the x coordinates of the ends of the moving rod at the same time, that is, simultaneously. The observer in S will find that L is shorter than the proper length L_{p} of the rod by an amount given by the **length contraction expression** $L = L_{\mathrm{p}}\sqrt{1 - v^2/c^2}$.

The **Doppler effect** for sound results from the motion of the source and/or observer through air, the medium in which the sound propagates. With electromagnetic waves there is no medium of propagation—there is no ether. As a consequence, the Doppler equations relating the frequency of light and other electromagnetic waves are different from the Doppler-effect relations for sound. Let f_0 be the frequency of a light source that is measured by an observer at rest with respect to the source. If this source moves toward another observer with a velocity v, the frequency f' that the latter observer measures is $f' = \sqrt{(1 + v/c)/(1 - v/c)}f_0$. If the source is moving away from an observer with a velocity v, the frequency f' that the observer measures is $f' = \sqrt{(1 - v/c)/(1 + v/c)}f_0$. In the latter case, $f' < f_0$, which is known as a **redshift**. This is most commonly observed in the light we see from distant receding galaxies.

Important Derived Results

γ-factor

$$\gamma = \frac{1}{\sqrt{1 - v^2/c^2}}$$

Lorentz transformations

$$x = \gamma(x' + vt') \qquad x' = \gamma(x - vt)$$
$$y = y' \qquad y' = y$$
$$z = z' \qquad z' = z$$
$$t = \gamma(t' + vx'/c^2) \qquad t' = \gamma(t - vx/c^2)$$

Galilean transformation

$$x = x' + vt' \qquad x' = x - vt$$
$$y = y' \qquad y' = y$$
$$z = z' \qquad z' = z$$
$$t = t' \qquad t' = t$$

Time dilation

$$\Delta t = \gamma\,\Delta t_{\mathrm{p}} = \frac{\Delta t_{\mathrm{p}}}{\sqrt{1 - v^2/c^2}}$$

Length contraction

$$L = \frac{1}{\gamma}L_{\mathrm{p}} = L_{\mathrm{p}}\sqrt{1 - v^2/c^2}$$

Relativistic Doppler effect, approaching

$$f' = \sqrt{\frac{1 + v/c}{1 - v/c}}f_0$$

Relativistic Doppler effect, receding

$$f' = \sqrt{\frac{1 - v/c}{1 + v/c}}f_0$$

Common Pitfalls

> Often you can answer questions about the space separation or time separation between two events by subtracting two appropriate Lorentz transformation equations from each other. The trick is to determine which are the "appropriate" Lorentz transformation equations to subtract. For example, if two events take place at the same location in the reference frame S', a

subtraction that involves $x'_2 - x'_1 = 0$ will probably prove useful in answering questions relating to the two events.

- Do not confuse the "time separation" between two events with the "proper time interval" between the two events. If an observer in reference frame S and a second observer in reference frame S' measure the time interval between two events that occur at different places for both observers, their time intervals will not be related by a simple multiplication or division by the factor $\sqrt{1 - v^2/c^2}$. To get the relationship between the time interval measured by each observer, you must use the previously discussed subtraction technique.
- Similarly, do not confuse the "spatial separation" between two events with "proper length." If an observer in reference frame S and a second observer in reference frame S' measure the spatial separation between two events that do not occur simultaneously, the spatial intervals will not be related by a simple multiplication or division by the factor $\sqrt{1 - v^2/c^2}$. To get the relationship between the spatial intervals measured by each observer, you must use the previously discussed subtraction technique.
- Often, problems can be solved using the simple expression distance = velocity × time, provided that the distance, velocity, and time all refer to the same reference frame. For example, suppose you are a muon moving at a speed $0.998c$ relative to Earth who has to traverse a distance of 9000 m measured by an observer on Earth. In your frame of reference, the distance to be traversed is foreshortened and is only $(9000\text{ m})\sqrt{1 - 0.998^2} = 600$ m, so the time Δt you would need to traverse this distance is found from

$$\begin{aligned} \text{distance} &= \text{velocity} \times \text{time} \\ 600\text{ m} &= 0.998(3.00 \times 10^8\text{ m/s})\Delta t \\ \Delta t &= 2.00 \times 10^{-6}\text{ s} \end{aligned}$$

3. TRUE or FALSE: A clock moving along the axis at constant speed is struck by a lightning bolt and is later struck by a second lightning bolt. The time interval recorded by this very sturdy clock between the two strikes is the proper time interval.

4. You find that it takes a time interval Δt for a spaceship to move through a distance L. How does the time interval elapsed on the clock of the pilot of the spaceship compare to what you measure? Explain.

Try It Yourself #1

An observer in reference frame S observes that a lightning bolt A strikes the x axis and 10^{-4} s later a second lightning bolt B strikes the x axis 1.50×10^5 m farther from the origin than A. What is the time separation between the two lightning bolts determined by a second observer in reference frame S' moving at a speed of $0.8c$ along the collinear x-x' axis?

Picture: Use the Lorentz transformation equation relating t, t' and x. Subtract the two times in the prime reference frame.

Solve:

Write an *algebraic* expression for the t' transformation.	
Subtract the expressions for events A and B to get the time difference in the prime reference frame.	$\Delta t' = -5.00 \times 10^{-4}$ s

Check: The time difference is still fairly small, which might be expected, and it is different from Δt in the S reference frame, which we should definitely expect.

Taking It Further: What, if anything, is the significance of having a negative time difference in the S' reference frame, when the time difference was positive in the S reference frame? Is it possible for an observer in reference frame S' to observe that bolts A and B strike at the same time? Explain.

Try It Yourself #2

A super rocket car traverses a straight track 2.40×10^5 m long in 10^{-3} s as measured by an observer next to the track. How much time elapses on a clock in the rocket car during this run?

Picture: Keep clear in your mind who is measuring each distance and time interval. Because the two clocks are moving with respect to one another, we will need to use time dilation. However, in order to determine the amount of time dilation, we need to know the relative speed of the clocks.

Solve:

Determine the speed of the car as measured by the observer at the side of the track. Express this result as a multiple of c.	

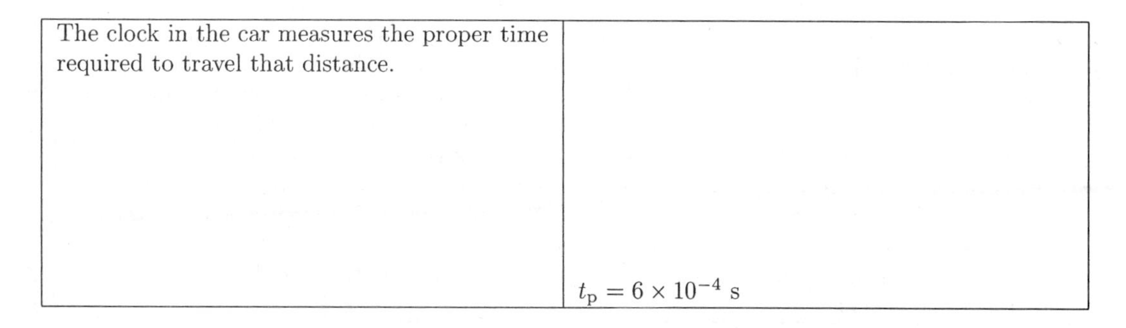

The clock in the car measures the proper time required to travel that distance.	$t_{\mathrm{p}} = 6 \times 10^{-4}$ s

Check: The proper time interval should always be shorter than any other measured time interval.

Taking It Further: What is the distance traveled in traversing the track as determined by the driver of the rocket car? Explain.

39.4 Clock Synchronization and Simultaneity

In a Nutshell

In a reference frame you generally do not want some of your clocks to be running ahead of or behind others. In a "good" reference frame, all clocks should be running together, that is, all clocks should be **synchronized** with one another. The Galilean transformation $t = t'$ tacitly assumes that it is obvious how to accomplish clock synchronization in all reference frames. If an observer in one reference frame adjusts his clocks to be synchronized, so that all clocks record the same time t for a given event, an observer in another reference frame also will "obviously" be able to adjust her clocks to be synchronized and measure the same time $t' = t$ for the event.

Einstein realized that an important quantity such as time should not be left imprecisely defined, as it was in Newtonian physics. Einstein pointed out that operational procedures must be used to define when two clocks are synchronous with each other. Einstein carefully defined clock synchronization using light signals. Clock synchronization is closely tied to the notion of simultaneity. If two events are **simultaneous**, it means that the two events occur at the same time in a given reference frame. The Galilean transformation $t = t'$ assumes that simultaneity is absolute. That is, it assumes that if one observer finds two events to be simultaneous, then all other observers moving relative to this observer will also find the two events to occur simultaneously.

Einstein defined simultaneity as follows: "Two events in a reference frame are simultaneous if light signals from the events reach an observer halfway between the events at the same time." Note that this definition involves an interweaving of the concepts of time, distance, and the speed of light.

A consequence of Einstein's operational definition of simultaneity is that two events that are simultaneous in one inertial reference frame will in general *not* be simultaneous in another inertial frame moving relative to the first. Since synchronization of clocks involves the notion of simultaneity, it also follows that two clocks that are synchronized in one reference frame will *not* be synchronized in any other frame moving relative to the first frame. If two clocks are synchronized in the frame in which they are both at rest, in a frame in which they are moving along the line through both clocks, the chasing clock leads (shows a later time than) the front clock by an amount $\Delta t_{\mathrm{s}} = L_{\mathrm{p}}(v/c)$ where L_{p} is the proper distance between the clocks.

One of the most famous results of time dilation is the **twin paradox**. One twin leaves her twin brother and takes a journey at a high constant speed v in a spaceship to a distant point, quickly turns around in a very small time interval, and returns to her twin brother at the same speed. When the twins compare their ages upon being reunited, they find that the brother has aged more than the sister according to the time dilation expression $\Delta t_{\text{brother}} = \Delta t_{\text{sister}}/\sqrt{1 - v^2/c^2}$. The "paradox" arises because, from the point of view of the twin sister, her brother moves away at the speed v, turns around, and then returns at the same speed. It therefore appears that the twin sister should find her brother to be younger—but you can't have it both ways. When the twins get together, one twin will definitely be younger than the other, or they will both have aged the same amount.

The paradox is resolved by noting that the motion of the two twins is not symmetrical. In order to return to her twin brother, the sister must turn around. The turning around is very real, resulting in the sister experiencing very real forces and accelerations during her turnaround period. She is equivalent to two observers in different inertial reference frames, one as she moves away and another as she moves toward her twin brother at the constant speed v. In contrast, the twin brother experiences no real net force nor acceleration throughout his twin's trip, and is equivalent to one single inertial observer. So the sister is, in fact, younger.

Common Pitfalls

- Since two events that occur simultaneously for one observer will not in general be simultaneous for other observers, you must make sure you are clear about which observer determines that the two events are simultaneous.
- Do not confuse the time at which you see an event take place with the time at which the event actually takes place. For example, suppose you see a distant supernova explode and record the time on a clock at your location when you make your observation. To determine the actual time that the supernova explosion occurred, you must correct for the travel time that it took the light signal to reach your location.

5. TRUE or FALSE: An observer in reference frame S finds that two lightning bolts strike the same place simultaneously. An observer in S' moving relative to S will determine that the two bolts strike at different times.

6. An observer in reference frame S' is moving relative to reference frame S at $0.8c$. At a certain instant a red flash is emitted at the 10.0-m mark on the x' axis of S' and 5.00 s later, as determined by clocks in S', a blue flash is emitted at the same 10.0-m mark. What is the proper time interval between the red and blue flashes? Explain.

Try It Yourself #3

Observers in reference frame S see an explosion located at $x_1 = 580$ m. A second explosion occurs 4.5 μs later at $x_2 = 1500$ m. In reference frame S', which is moving along the $+x$ axis at speed v, the explosions occur at the same point in space. What is the separation in time between the two explosions as measured in S'?

Picture: Since the explosions occur at the same point in space in S', $\Delta t'$ is the proper time. You will need to use the location of the two explosions in S, as well as their temporal separation, to find the speed of the two reference frames relative to each other.

Solve:

Write an *algebraic* expression for the time interval $\Delta t'$.	
Use the spatial separation of the explosions in reference frame S and the measured time interval in S to find the relative speed of S and S'.	
Substitute the speed calculated above, and all other given quantities, with their units, into the expression of the first step, and solve for the time interval $\Delta t'$.	$\Delta t' = \Delta t_{\mathrm{p}} = 3.29\ \mu\mathrm{s}$

Check: Since the events occur at the same spatial location in S', the time interval $\Delta t'$ is the proper time. So it should be smaller than Δt, which it is.

Try It Yourself #4

A spaceship departs from Earth for the star Alpha Centauri, which is 4 light-years away. The spaceship travels at $0.77c$. What is the time required to get there as measured by a passenger on the spaceship?

Picture: Passengers on the spaceship measure a contracted distance relative to the distance measured from Earth. The time required is the time needed to travel this contracted distance.

Solve:

Determine the contracted distance between Earth and Alpha Centauri observed by a passenger on the spaceship.	

Determine the time required to travel this distance given the speed of the spaceship.	$\Delta t = 3.31$ y

Check: Because the arrival and departure events both occur in the spaceship, the time measured by the spaceship should be the proper time, and hence shorter than the time measured by an earthbound observer, which it is.

Taking It Further: How long does it take for the spaceship to arrive at Alpha Centauri as measured on Earth?

39.5 The Velocity Transformation

In a Nutshell

Suppose observers in reference frames S and S' measure the velocity of a single object, such as a spaceship. The observer in S measures the three components of the velocity to be (u_x, u_y, u_z) whereas the observer in S' measures in general three different components of velocity (u'_x, u'_y, u'_z). By differentiating the Lorentz transformation equations, the **relativistic velocity transformation equations**, shown at the right, relating the velocity components are obtained. In these equations v is the speed of frame S' relative to S in the positive x direction.	$u_x = \dfrac{u'_x + v}{1 + vu'_x/c^2}$ $u_y = \dfrac{u'_y}{\gamma(1 + vu'_x/c^2)}$ $u_z = \dfrac{u'_z}{\gamma(1 + vu'_x/c^2)}$	$u'_x = \dfrac{u_x - v}{1 - vu'_x/c^2}$ $u'_y = \dfrac{u_y}{\gamma(1 - vu'_x/c^2)}$ $u'_z = \dfrac{u_z}{\gamma(1 - vu'_x/c^2)}$
For low velocities ($v \ll c$) the relativistic velocity transformation equations reduce to the classical velocity transformation equations.	$u_x = u'_x + v$ $u_y = u'_y$ $u_z = u'_z$	$u'_x = u_x - v$ $u'_y = u_y$ $u'_z = u_z$

Important Derived Results

Relativistic velocity transformation

$$u_x = \frac{u'_x + v}{1 + vu'_x/c^2} \qquad u'_x = \frac{u_x - v}{1 - vu'_x/c^2}$$

$$u_y = \frac{u'_y}{\gamma\,(1 + vu'_x/c^2)} \qquad u'_y = \frac{u_y}{\gamma\,(1 - vu'_x/c^2)}$$

$$u_z = \frac{u'_z}{\gamma\,(1 + vu'_x/c^2)} \qquad u'_z = \frac{u_z}{\gamma\,(1 - vu'_x/c^2)}$$

Classical velocity transformation

$$u_x = u'_x + v \qquad u'_x = u_x - v$$

$$u_y = u'_y \qquad u'_y = u_y$$

$$u_z = u'_z \qquad u'_z = u_z$$

Common Pitfalls

> Velocity problems usually involve two reference frames S and S' and a particle P. In the velocity transformation expressions, the quantity v is the velocity of reference frame S' relative to reference frame S. The quantities (u_x, u_y, u_z) and (u'_x, u'_y, u'_z) are the components of the particle's velocity relative to reference frames S and S' respectively. Don't confuse v with u or u'. Also, make sure you clearly understand which objects are to be associated with S, S', and P .

7. TRUE or FALSE: Rocket ship A moves away from you to your right with a speed $0.5c$, while rocket ship B moves away from you toward your left with a speed $0.4c$. The speed of A as determined by B is $0.9c$.

8. Triplets A and B take trips in high-speed spaceships while triplet C stays at home. Triplet A moves along the positive x axis to a far galaxy and then returns home at the same speed. Triplet B moves along the negative x axis at the same speed as A to the same distance as A, and then returns home at the same speed. Compare the ages of the triplets when they are all together again.

Try It Yourself #5

A radioactive nucleus moving at a speed of $0.8c$ in a laboratory decays and emits an electron in the same direction as the nucleus is moving. The electron moves at $0.6c$ relative to the nucleus. How fast is the electron moving according to an observer in the laboratory?

Picture: Associate the objects in the problem with observers in reference frames S and S' and particle P. Determine which velocities correspond to v, u, and u' and apply the appropriate velocity transformation equations.

Solve:

Make the appropriate reference frame assignments.	
Make the corresponding velocity assignments.	
Apply the velocity transformation equation.	speed of electron in lab frame $= u_x = 0.946c$

Check: Our intuition tells us that we should expect the speed to be greater than $0.8c$. However, from relativity we know that the speed must also be less than c. The speed we found is within that range.

Try It Yourself #6

Rocket A travels away from Earth at $0.6c$, and rocket B travels away from Earth in exactly the opposite direction at $0.8c$. What is the speed of rocket B as measured by the pilot of rocket A?

Picture: Associate the objects in the problem with observers in reference frames S and S' and particle P. Determine which velocities correspond to v, u, and u' and apply the appropriate velocity transformation equations.

Solve:

Make the appropriate reference frame assignments.	
Make the corresponding velocity assignments.	
Use the velocity transformation equation.	$v_{\text{B measured by A}} = 0.946c$

Check: Our intuition tells us that we should expect the speed to be greater than $0.8c$. However, from relativity we know that the speed must also be less than c. The speed we found is within that range.

Taking It Further: Why did we get the same answer for this problem that we did for the previous problem?

39.6 Relativistic Momentum

In a Nutshell

If an object is at rest with respect to an inertial observer, the observer can measure its mass in a straightforward manner, for instance by putting it on a balance. This mass is the **rest mass** of the object. This mass is sometimes written as m_0. For the purposes of this text, we will treat the terms *mass* and *rest mass* as synonymous, and we will label them simply m.

The **relativistic momentum** $\vec{p}$ of an object moving with a velocity $\vec{u}$ is the given by $\vec{p} = m\vec{u}/\sqrt{1 - u^2/c^2}$. This quantity is conserved when there are no net external forces, and for small velocities $u \ll c$, this reduces to the usual classical momentum expression $\vec{p} = m\vec{u}$.

Important Derived Results

Relativistic momentum $$\vec{p} = \frac{m\vec{u}}{\sqrt{1-u^2/c^2}}$$

Common Pitfalls

> Do not mistakenly use the expression $\vec{p} = m\vec{u}$ for an object's momentum in relativistic situations. Instead use the correct relativistic expression given above.

9. TRUE or FALSE: A consequence of the relativistic velocity transformation is that if one observer measures something to move at a speed c in an arbitrary direction, then an observer in any other reference frame will also measure the speed of the object to be c.

39.7 Relativistic Energy

In a Nutshell

It follows from a relativistic analysis of work–energy principles that a particle with rest mass m has a rest energy E_0 given by Einstein's famous **mass-energy relationship**: $E_0 = mc^2$. If the particle is moving with a speed u, so that it has a kinetic energy K, its **total relativistic energy** $E = K + E_0$ is defined as $E = K + E_0 = K + mc^2 = mc^2/\sqrt{1-u^2/c^2}$. If the object is not moving, $u = 0$ and $E = E_0 = mc^2$.

Turning around the expression for the total relativistic energy, you find that the **relativistic kinetic energy**—the energy due to the particle's motion—is the difference between the total energy E and the rest energy E_0: $K = E - E_0 = mc^2/\sqrt{1-u^2/c^2} - mc^2$. For small velocities ($u \ll c$) application of the binomial expansion shows that the relativistic kinetic energy reduces to the usual standard classical expression $K_{\text{classical}} = \frac{1}{2}mu^2$.

The relativistic momentum and relativistic energy expressions are related. A useful expression for the speed u of a particle is $u/c = pc/E$. When the speed u is eliminated between the momentum and energy relations, the following expression relating the energy and momentum is obtained: $E^2 = p^2c^2 + (mc^2)^2$, or, since $E = K + mc^2$, $(K + mc^2)^2 = p^2c^2 + (mc^2)^2$. If you know a particle's momentum, these expressions allow you to find its total energy E or kinetic energy K, and vice versa.

The constituent parts of an atom or a nucleus, or of any other particle, have less potential energy when together than when separated. The difference is called the **binding energy** (E_b) of the system. The expression $E_0 = mc^2$ shows that binding energy is equivalent to mass. As a result, the rest mass of a composite particle will be less than the rest masses of its constituent parts by an amount equal to the binding energy.

Physical Quantities and Their Units

Rest energy of electron	$m_{\text{electron}}c^2 = 0.511$ MeV
Rest energy of proton	$m_{\text{proton}}c^2 = 938.3$ MeV
Rest energy of neutron	$m_{\text{neutron}}c^2 = 939.6$ MeV

Fundamental Equations

Rest energy	$E_0 = mc^2$

Important Derived Results

Total relativistic energy	$E = K + mc^2 = \dfrac{mc^2}{\sqrt{1 - u^2/c^2}}$
Relativistic kinetic energy	$K = E - E_0 = \dfrac{mc^2}{\sqrt{1 - u^2/c^2}} - mc^2$
Relativistic velocity-momentum-energy relationship	$\dfrac{u}{c} = \dfrac{pc}{E}$
Relativistic energy-momentum relationships	$E^2 = \left(K + mc^2\right)^2 = p^2c^2 + \left(mc^2\right)^2$

Common Pitfalls

> Often particles are described in terms of their kinetic energies rather than their speeds. For example, you may be told that a particle of charge e has been accelerated from rest through a certain voltage, which is numerically equal to the kinetic energy in electron volts that the particle acquires. Relativistic expressions must be used when the kinetic energy is comparable to the rest energy of an object. For example, you must use relativistic expressions for a 2-MeV electron, because an electron's rest mass is about 0.511 MeV. For a 2-MeV proton, however, you can use nonrelativistic expressions to a good approximation because a proton's rest mass is about 938 MeV.

10. TRUE or FALSE: The kinetic energy of a proton that has been accelerated from rest through a potential difference of 500×10^6 V is 500 MeV.
11. In terms of energies, when can you use $p = mu$ to a good approximation? Explain.

Try It Yourself #7

How fast is a proton with $K = 1\,000\,000$ MeV moving?

Picture: If $K \ll mc^2$, then you can use the nonrelativistic expression for kinetic energy. Otherwise, you must use the relativistic expression for kinetic energy to determine the proton's velocity.

Solve:

Look up the rest mass of the proton and compare to the kinetic energy given to determine which expression to use.	
Determine the speed from the appropriate kinetic energy expression.	$v = 0.999999c$

Check: The speed is less than the speed of light, as expected.

Taking It Further: If you used the nonrelativistic expression for kinetic energy, what speed would you find? How do you know this is wrong?

Try It Yourself #8

What is the kinetic energy of an electron whose momentum is 300 MeV/c?

Picture: If $pc \gg mc^2$, then you can use $E = pc$ or $K = pc$ to a good approximation. Otherwise, you will need to use the relativistic expression relating energy and momentum to find the electron's energy.

Solve:

Look up the rest mass of the electron and compare to the momentum given to determine which expression to use.	
Evaluate the kinetic energy using the appropriate expression.	$K = 300$ MeV

Check: Since the rest energy of an electron is only 0.511 MeV, the total energy pc for this problem is essentially all kinetic energy.

Taking It Further: How would this change if the particle were a proton?

39.8 General Relativity

In a Nutshell

Einstein's **general theory of relativity** is a theory of gravitation that supersedes Newton's theory of gravitation. Newton's theory of gravitation states that the magnitude of the gravitational attractive force F_g between two massive bodies m_1 and m_2 separated by a distance R is $F_g = Gm_1m_2/R^2$. General relativity describes what happens in the realm of very strong gravitational fields where Newton's theory of gravitation no longer holds.

Einstein was guided to his general theory of relativity by the **principle of equivalence**. In brief, the principle of equivalence states that in a small region of space and a short interval of time you cannot distinguish between real gravitational fields and accelerated motion: "A homogeneous gravitational field is completely equivalent to a uniformly accelerated reference frame." If you are in a spaceship undergoing uniform acceleration, any experiment you perform will be exactly the same as if you were at rest in a uniform gravitational field. For example, if you release an object in the spaceship, it moves away from you with constant acceleration in the same manner that an object moves away from you when you release it in a gravitational field, as shown.

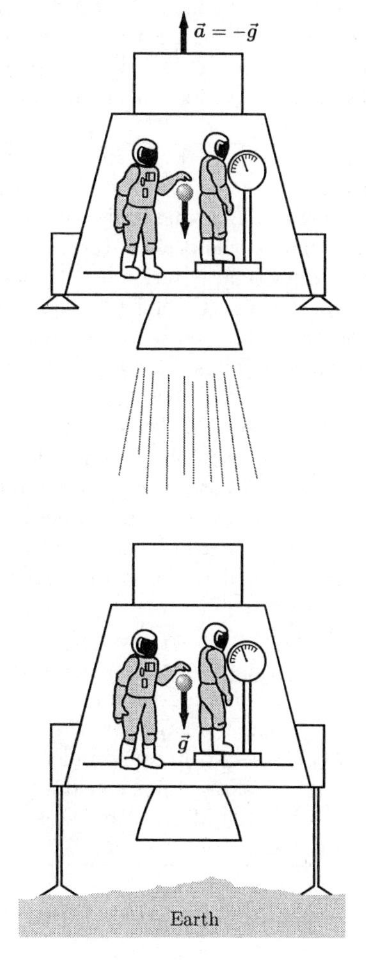

One consequence of the principle of equivalence is that light will follow a curved trajectory in a gravitational field, since light follows such a trajectory in an accelerated reference frame. The bending of a light beam as observed in an accelerated elevator is illustrated in the figure below. Figure (a) shows the light beam moving in a straight line, while the elevator accelerates upward. Figure (b) shows how the light beam follows a curved trajectory when viewed in the accelerating elevator. Since the light beam is curved in the accelerated elevator, the principle of equivalence states that light should also bend in a gravitational field.

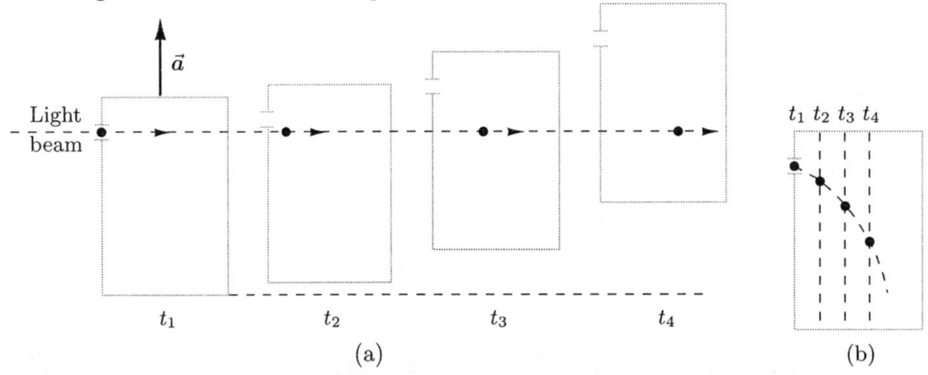

(a) (b)

The bending of light in a gravitational field was first observed in 1919, about three years after Einstein published his theory of general relativity, by a team of scientists led by Sir Arthur Eddington, who made observations of the bending of light during an eclipse of the sun that agreed exactly with Einstein's prediction. This observation immediately catapulted Einstein into worldwide fame.

Another prediction of Einstein's theory of general relativity is that clocks in a region of low gravitational potential will run slower than clocks in a region of higher gravitational potential. This slowing down of time is also often referred to as the gravitational redshift. An atomic clock measures time on the basis of the frequency of vibrations of an atom. If an atomic clock moves to a region of low gravitational potential, the frequency of this vibration slows down, resulting in a longer wavelength of light being emitted by the atom. This slowing down of time was experimentally observed with an atomic clock in 1976. Geographical positioning system (GPS) receivers and satellites must account for this gravitational redshift in order to provide accurate position data. If the gravitational affect were not accounted for, the resulting errors in position would make the system useless in less than one day.

Yet another prediction of Einstein's theory of general relativity is the possible existence of **black holes**. The concept of a black hole is built around the **Schwarzschild radius** R_S associated with the mass M of an astronomical object: $R_S = 2GM/c^2$, where G is Newton's gravitational constant and c is the speed of light. Given the mass M of the astronomical object, its Schwarzschild radius can be calculated from this expression.

If an astronomical object with a given mass is sufficiently dense, it is possible for the Schwarzschild radius calculated from its mass to be larger than the radius of the object. For such a situation, the present standard view is that the region extending from immediately outside the surface of the object to the Schwarzschild radius lies in a black hole, and that light emitted from the surface of the object cannot escape from the black hole to the outer region. In principle, a black hole can arise from an object collapsing under the influence of gravitational forces to smaller and smaller radii, while its mass remains constant, until a black hole is formed around the collapsing object. At the center of the Milky Way is a supermassive black hole with a mass of two to four million solar masses.

Physical Quantities and Their Units

Schwarzschild radius $R_S = 2GM/c^2$

QUIZ

1. TRUE or FALSE: All observers in inertial reference frames will measure the speed of light in any direction to be 3.00×10^8 m/s.

2. TRUE or FALSE: An electron and a proton are each accelerated through a potential difference of 50,000 V and are then injected into the magnetic field of a cyclotron. You must use relativistic expressions to analyze the motion of the electron, but the motion of the proton can be treated with classical expressions to a good approximation.

3. Explain how the measurement of time enters into the determination of the length of an object.

4. Is it possible for one observer to find that event A happens after event B and another observer to find that event A happens before event B? Explain.

5. In terms of energies, when can you use $p = E/c$ to a good approximation? Explain.

6. What is the kinetic energy of a proton whose momentum is 500 MeV/c?

7. A beam of pions has a speed of $0.88c$. Their mean lifetime, as measured in the reference frame of the laboratory, is 2.6×10^{-8} s. What is the distance traveled by the laboratory as measured by the pion, during its lifetime?

Chapter 40

Nuclear Physics

40.1 Properties of Nuclei

In a Nutshell

Nuclei are composed of positively charged protons and uncharged neutrons, referred to collectively as nucleons; their main properties are given below.

Property	Proton	Neutron
Charge	$+1.602 \times 10^{-19}$ C	0
Rest Mass	1.672623×10^{-27} kg	1.674929×10^{-27} kg
	$938.2723\ \mathrm{MeV}/c^2$	$939.5656\ \mathrm{MeV}/c^2$
	1.007277 u	1.008665 u
Spin	$\frac{1}{2}$	$\frac{1}{2}$

The numbers of protons and neutrons in a nucleus are designated as follows:

Z = number of protons, called the **atomic number**

N = number of neutrons

$A = N + Z$ = number of nucleons, called the **mass number**

A particular nuclear species, called a **nuclide**, is designated by giving the atomic symbol X of the nucleus, with the mass number A as a presuperscript and the atomic number Z as a presubscript, in the form ${}^{A}_{Z}X$. Often the presubscript Z is not used because it is associated uniquely with the atomic symbol X.

Isotopes are nuclides with the same atomic number Z but different numbers N of neutrons (and correspondingly different mass numbers A). For example, ${}^{12}_{6}\mathrm{C}$, ${}^{13}_{6}\mathrm{C}$ and ${}^{14}_{6}\mathrm{C}$ are all isotopes of carbon.

Nucleons in a nucleus attract one another with a **strong nuclear force** (also called a **hadronic force**). Because this attractive force is independent of the charge of the nucleons, proton-proton, proton-neutron, and neutron-neutron strong nuclear forces are all roughly equal. The nuclear force has a short range, being essentially zero between nucleons separated by more than a few femtometers (10^{-15} m).

To a very good approximation a nucleus can be treated as a sphere of radius R given by $R = R_0 A^{1/3}$, where R_0 is about 1.2 fm. The value of R_0 depends on which nuclear property is being measured: charge distribution, mass distribution, region of influence of the strong nuclear force, and so forth. Because the mass and volume of a nucleus are both proportional to A, the densities of all nuclei are approximately the same.

For light stable nuclei the number of protons Z and neutrons N are about equal to each other. As the number of nucleons increases, N becomes larger than Z for stable nuclei, as shown in the plot of N versus Z.

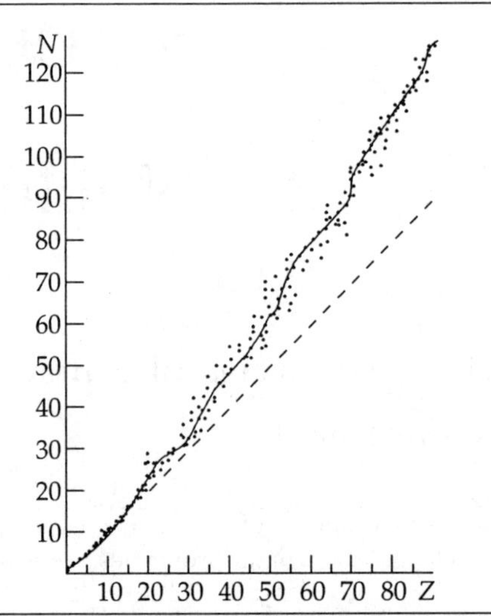

The rest mass of a stable nucleus is less than the sum of the rest masses of its constituent nucleons because of the energy required to bind the nucleons together (see Section 39-6). The **binding energy** of a nucleus of atomic mass can be written as $E_b = (ZM_H + Nm_n - M_A)c^2$, where M_H is the mass of a ^{1}H atom and m_n is the mass of a neutron. The reason for using atomic masses for ^{1}H and M_A instead of nuclear masses is that it is the atomic masses that are directly measured and listed in tables. The mass of the extra Z electrons in the ZM_H term is canceled by the mass of the extra Z electrons in the M_A term.

A quantity of interest is the **average binding energy per nucleon**, E_b/A, obtained by dividing the binding energy of a nucleus by the number of nucleons in the nucleus. A plot of E_b/A versus the mass number A is shown. For $A > 50$, E_b/A is roughly constant at about 8 MeV per nucleon, indicating that E_b is approximately proportional to the number of nucleons A in a nucleus.

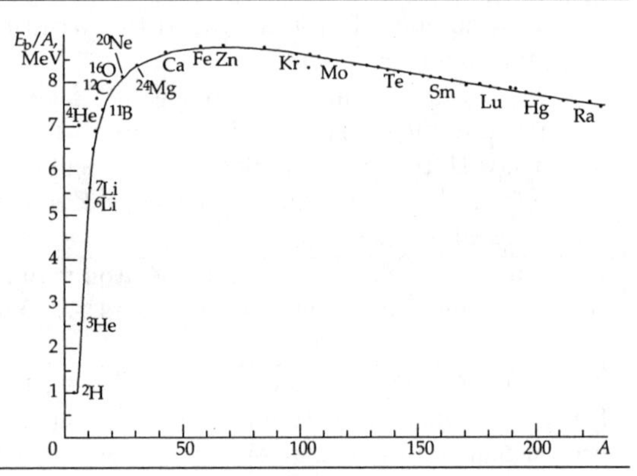

Physical Quantities and Their Units

Rest mass of a proton	1.672623×10^{-27} kg $= 938.2723$ MeV$/c^2 = 1.007277$ u
Rest mass of a neutron	1.674929×10^{-27} kg $= 939.5656$ MeV$/c^2 = 1.008665$ u

Important Derived Results

Nuclear radius	$R = R_0 A^{1/3}$
Binding energy of a nucleus	$E_b = (ZM_H + Nm_n - M_A)c^2$

Common Pitfalls

- In the expression for the binding of energy of a nucleus, do not mistakenly use the mass of a proton instead of the atomic mass of hydrogen. Make sure you understand how the use of atomic masses (for which there are tables) instead of nuclear masses is justified: the masses of the electrons cancel.
- Do not confuse the total binding energy of a nucleus with the binding energy per nucleon of the nucleus.
- Understand that the average binding energy per nucleon is different from the energy required to remove a single nucleon from a nucleus.

1. TRUE or FALSE: A nucleus is approximately a sphere with constant density.
2. What is an isotope?

Try It Yourself #1

Given that a nucleus is approximately spherical, with a radius $R = R_0 A^{1/3}$, where R_0 is about 1.2 fm, determine its approximate mass density. Express your answer in SI units and also in tons per cubic inch.

Picture: Density is mass per unit volume.

Solve:

Find an *algebraic* expression for the mass of the nucleus.	
Find an *algebraic* expression for the volume of the nucleus.	
Determine the density of the nucleus.	density $= 2.35 \times 10^{17}$ kg/m^3 $= 4.20 \times 10^{9}$ ton/in^3

Check: This mass density is incredibly large in comparison to the density of materials we experience in everyday life. However, nuclei are pretty dense.

Taking It Further: Explain why nuclei are so much more dense than materials we experience in everyday life.

Try It Yourself #2

What is the binding energy of a ^{12}C nucleus?

Picture: $E_b = (ZM_H + Nm_n - M_A)c^2$.

Solve:

Determine the number of protons and neutrons in a ^{12}C nucleus.	
Determine the atomic mass of ^{12}C.	
Find the binding energy.	$E_b = 92.2$ MeV

Check: This energy is about 0.1 times the energy of a single nucleon, so it seems reasonable.

Taking It Further: What would it mean if the binding energy were negative?

40.2 Radioactivity

In a Nutshell

An unstable nucleus decays into another nucleus accompanied by the emission of a **decay product**. The **radioactive decay** is characterized by the decay product according to the nomenclature shown in the table.

Type of decay	Decay product
Alpha (α) decay	^{4}He nuclei (alpha particles)
Beta (β) decay	Electrons: e^- or β^-; or positrons: e^+ or β^+ (beta particles)
Gamma (γ) decay	Photons (gamma rays)

Radioactive decay is a statistical process, and as such it follows an exponential decay dependence. If there are N_0 radioactive nuclei at time zero, the number N of radioactive nuclei remaining after a time t is given by $N = N_0\,\mathrm{e}^{-\lambda t}$, where the constant λ, called the **decay constant**, depends on the particular nucleus and the decay process taking place.

The reciprocal of the decay constant is the **average** or **mean lifetime** τ: $\tau = 1/\lambda$. At the end of a mean lifetime, a radioactive sample has decayed to $1/\mathrm{e}$ or 37 percent of the original number of nuclei. Another time that characterizes decay rate is the **half-life** $t_{1/2}$, which is the time it takes for the number of radioactive nuclei in a sample to decrease to one-half the original number. As shown, if you start with N_0 nuclei, after one half-life $N_0/2$ nuclei will remain, after two half-lives $N_0/4$ nuclei will remain, and so forth. The half-life is related to the decay constant λ by $t_{1/2} = \ln 2/\lambda = 0.693/\lambda = 0.693\tau$.

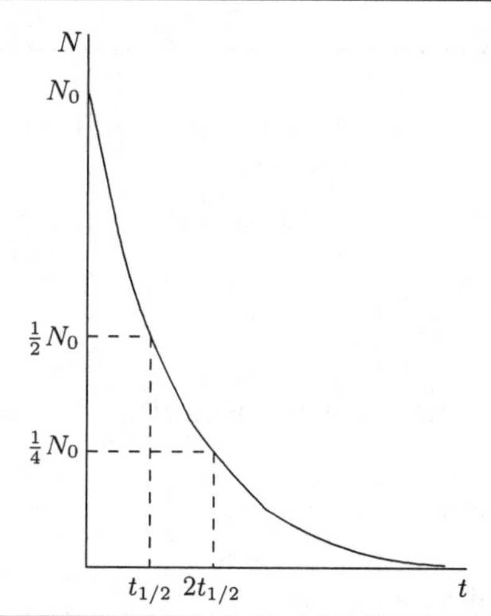

The number of decays per second, called the **decay rate** R, is found by differentiating the expression for N to get $R = -dN/dt = \lambda N = \lambda N_0\,\mathrm{e}^{-\lambda t} = R_0\,\mathrm{e}^{-\lambda t}$, where $R_0 = \lambda N_0$ is the decay rate at $t = 0$. The SI unit of measure of decay rate is the **becquerel** (Bq), defined as one decay per second: 1 Bq = 1 decay/s. Another common unit by which decay rates are measured is the **curie** (Ci), defined as 1 Ci = 3.7×10^{10} decays/s = 3.7×10^{10} Bq. A curie is approximately equal to the rate of decay of 1.00 g of radium. After a time equal to one half-life, the decay rate decreases by a factor of 2.

In every type of radioactive decay, a **parent nucleus** emits a decay product and decays into a **daughter nucleus**. Essentially, the birth of a daughter results from the death of the parent. We will look at three types of decay in more detail: beta, gamma, and alpha decay. In each decay the laws of conservation of mass-energy, charge, and linear and angular momentum hold. In addition, there is a law of conservation of nucleons, according to which the number of nucleons after a decay equals the number of nucleons before the decay.

In **gamma (γ) decay**, a nucleus in an excited state emits a photon, called a γ ray, as it makes a transition to a lower energy state, analogous to an atom in an excited state emitting a photon as it decays. The gamma decay reaction can be written as $\left({}^A_Z X\right)^* \to {}^A_Z X + \gamma$, where the $\star$ denotes an excited state of the nucleus. In γ decay the parent and daughter are the same nucleus. The γ rays are observed to be emitted with discrete wavelengths, showing that nuclei possess discrete energy levels typically separated by energies on the order of MeV.

In **alpha** (α) **decay** a parent nucleus decays into a daughter nucleus with the emission of an α particle (^{4}He nucleus): ${}^A_ZP \rightarrow {}^{A-4}_{Z-2}D + {}^4_2\text{He}$. Thus in α decay, N and Z both decrease by 2, and A decreases by 4.

In $\boldsymbol{\beta^-}$ **decay**, a parent P decays into a daughter D by conversion of a neutron into a proton with the emission of an electron (β^-) and an antineutrino: ${}^A_ZP \rightarrow {}_{Z+1}^{\ \ A}D + \beta^- + \bar{\nu}_e$. In $\boldsymbol{\beta^+}$ **decay**, a proton is converted into a neutron with the emission of a positron (β^+) and a neutrino: ${}^A_ZP \rightarrow {}_{Z-1}^{\ \ A}D + \beta^+ + \nu_e$. The electron-associated **neutrino** (ν_e) or **antineutrino** ($\bar{\nu}_e$) that appears as one of the decay products was initially thought to have zero rest mass, zero charge, spin $\frac{1}{2}$ and to move at the speed of light. It is now believed that the neutrino has some very small, but finite, mass. If the neutrino or antineutrino were not emitted as a decay product in β decay, energy, momentum, and spin would not be conserved. In fact, it was consideration of conservation of energy and momentum in β decay that motivated Pauli in 1930 to postulate the existence of a neutrino particle. Eventually, in 1956, a neutrino was observed experimentally by Cowan and Reines. The distinction between the various types of neutrinos will be explained in Chapter 41.

In many cases the daughter of a radioactive nucleus is also unstable and decays further by α or β decay.

Physical Quantities and Their Units

Definition of SI unit of decay rate	$1\ \text{Bq} = 1\ \text{decay/s}$
Definition of the curie	$1\ \text{Ci} = 3.7 \times 10^{10}\ \text{decays/s} = 3.7 \times 10^{10}\ \text{Bq}$

Important Derived Results

Radioactive decay law	$N = N_0\,\mathrm{e}^{-\lambda t}$
Mean lifetime	$\tau = \dfrac{1}{\lambda}$
Half-life	$t_{1/2} = \dfrac{\ln 2}{\lambda} = \dfrac{0.693}{\lambda} = 0.693\tau$
Radioactive decay rate	$R = -\dfrac{dN}{dt} = \lambda N = \lambda N_0\,\mathrm{e}^{-\lambda t} = R_0\,\mathrm{e}^{-\lambda t}$
β^- decay	${}^A_ZP \rightarrow {}_{Z+1}^{\ \ A}D + \beta^- + \bar{\nu}_e$
β^+ decay	${}^A_ZP \rightarrow {}_{Z-1}^{\ \ A}D + \beta^+ + \nu_e$
α decay	${}^A_ZP \rightarrow {}^{A-4}_{Z-2}D + {}^4_2\text{He}$
γ decay	$\left({}^A_ZX\right)^* \rightarrow {}^A_ZX + \gamma$

Common Pitfalls

- Understand that the symbols β^- and e^- are used interchangeably for electrons, as are the symbols β^+ and e^+ for positrons.
- Do not think that the electrons or positrons emitted in β decay are contained inside a nucleus. They are not. Beta particles are created in the process of decay much as photons, which do

not reside in an atom, are created when an atom makes a transition from a higher to a lower energy level. The same holds for alpha decay and gamma decay.

3. TRUE or FALSE: After β^- decay a nucleus has one more proton and one fewer neutron than before the decay.

4. Explain how conservation of energy and momentum would be violated if a neutrino were not emitted in beta decay.

Try It Yourself #3

The curie unit is defined as 1 Ci = 3.7×10^{10} Bq, which is about the rate at which radiation is emitted by 1.00 g of radium. Calculate the half-life of radium from this definition.

Picture: Avogadro's number can be used to find the number of atoms. Use this to find the decay constant, which can be used to determine the half-life.

Solve:

Find the number of atoms in 1.00 g of radium.	
Find the decay constant from the number of atoms and the decay rate.	
Determine the half-life.	$t_{1/2} = 1580$ y

Check: The units work out.

Taking It Further: The actual measured half-life of radium is 1620 y. What does that, and your calculation, tell you about the radiation emission rate of radium?

Try It Yourself #4

The decay rate of a sample of radioactive $^{200}_{79}\text{Au}$ is measured as 15.0 mCi and 10 min. later as 13.0 mCi. Determine the half-life of $^{200}_{79}\text{Au}$.

Picture: Find the decay constant from the expression for the decay rate. Use the decay constant to determine the half-life.

Solve:

Determine the decay constant from the two decay rates separated by time.	
Use this decay constant to find the half-life.	$t_{1/2} = 48.4$ min

Check: The units work out properly.

40.3 Nuclear Reactions

In a Nutshell

Information about a nucleus can be obtained by bombarding it with a known projectile and analyzing the resulting nuclear reaction. Many types of nuclear reactions can take place. In one typical nuclear reaction, after the projectile interacts with the target, one or more particles are detected with experimental apparatus and a residual nucleus is left unobserved: target nucleus + projectile → undetected residual nucleus + detected particle(s). Sometimes the residual nucleus is radioactive and decays by the emission of other particles that can be detected. In any nuclear reaction equation, the total charge Z and total number of nucleons A must be the same on both sides of the equation.

The ***Q* value** of a nuclear reaction is the difference between the total rest masses before and after the reaction: $Q = \left(\sum m_0c^2\right)_{\text{before}} - \left(\sum m_0c^2\right)_{\text{after}} = -(\Delta m)c^2$. The Q value can be positive or negative. For an **exothermic reaction** the Q value is positive, and energy is released in the reaction. An exothermic reaction can occur even when both initial particles are at rest. In an **endothermic reaction** the Q value is negative, and energy is absorbed in the reaction. An endothermic reaction cannot occur unless the bombarding particle has a kinetic energy greater than a certain threshold value because of conservation of energy.

A **cross section** σ is a measure of the probability that a bombarding particle will interact with the target nucleus to produce a specific nuclear reaction. The cross section for a reaction is defined as the ratio of the number of reactions per second per nucleus, R, to the number of projectiles incident per second per unit area, I (the incident intensity): $\sigma = R/I$. The cross section has the dimensions of area and is commonly measured in a unit called the barn: 1 barn = 10^{-28} m^2, which is of the order of the square of a nuclear radius. The larger the value of σ, the more likely a reaction will occur. A nuclear cross section of 1 barn is relatively large, so it is said that a nuclear target with a cross section of 1 barn is as easy to hit as the "broad side of a barn."

Physical Quantities and Their Units

Unit of cross section	1 barn = 10^{-28} m^2

Important Derived Results

Q value	$Q = \left(\sum m_0c^2\right)_{\text{before}} - \left(\sum m_0c^2\right)_{\text{after}} = -(\Delta m)c^2$
Cross section	$\sigma = \dfrac{R}{I}$

Common Pitfalls

> Keep straight the difference between an exothermic and an endothermic reaction. The former emits energy; the latter absorbs it.

5. TRUE or FALSE: The cross section for an endothermic reaction is zero below the reaction's threshold energy.

6. What are important physical factors that can influence the cross section of a reaction, and what are their effects?

Try It Yourself #5

How much energy in MeV is released or absorbed in the reaction $^{150}_{62}\text{Sm} + \text{p} \rightarrow {}^{147}_{61}\text{Pm} + \alpha$, given the atomic masses $^{150}_{62}\text{Sm} = 149.917276$ u and $^{147}_{61}\text{Pm} = 146.915108$ u.

Picture: Determine the Q value. Although this is a nuclear reaction, atomic masses are used, so we must conserve the number of electrons as well.

Solve:

Write an *algebraic* expression for the Q value of the nuclear reaction in terms of atomic masses.	

Look up the atomic masses of hydrogen and helium, the atoms that correspond to the proton and alpha particle. Use as many significant digits as possible.	
Substitute values for the atomic masses into the expression for the Q value and solve.	$Q = +6.96$ MeV

Check: The units work out, and this energy seems reasonable. It is much less than the rest energy of a single proton or neutron.

Taking It Further: Is this reaction endothermic or exothermic? Explain.

Try It Yourself #6

Assume in a typical alpha decay that the parent with A nucleons decays while at rest. In the approximation that the mass of each nucleus in the reaction is proportional to its number of nucleons, determine the kinetic energies of the alpha particle and daughter nucleus in terms of the original A number and the Q value of the reaction. Use nonrelativistic calculations.

Picture: Energy and momentum must be conserved. The Q value will give the sum of the kinetic energies of the alpha particle and daughter nucleus.

Solve:

Write a generic expression for the alpha decay described.	

Assuming the masses of the particles are proportional to the number of nucleons, determine the ratio of alpha mass to the daughter mass.	
Conserve momentum. The magnitudes of the daughter and alpha momenta must be the same.	
Find the ratio of the kinetic energies of the particles. We can use nonrelativistic expressions.	
Determine the Q value of the reaction, which is the total kinetic energy of the particles after the reaction.	
Solve the above expression for the kinetic energy of the alpha particle in terms of Q and A.	$K_\alpha = \frac{A-4}{A}Q$
Use the expressions in steps 5 and 6 to solve for the kinetic energy of the daughter particle. Since for a given alpha decay reaction the Q value is fixed, the alpha particle comes off with a precise kinetic energy, that is, it is monoenergetic. The larger the value of A the smaller the daughter's kinetic energy, and the kinetic energy of the alpha particle is more nearly equal to the Q value of the reaction.	$K_D = \frac{4Q}{A}$

Check: $K_\alpha + K_D = Q$, which it should.

40.4 Fission and Fusion

In a Nutshell

The figure shows a cartoon of a **typical nuclear fission reaction**: (a) a heavy nucleus ($A > 200$) such as ${}^{235}_{92}\text{U}$ absorbs a neutron, producing (b) an intermediate nucleus that is in an excited state; the excited intermediate nucleus (c) then splits into two medium-mass nuclei accompanied by the emission of several neutrons (d).

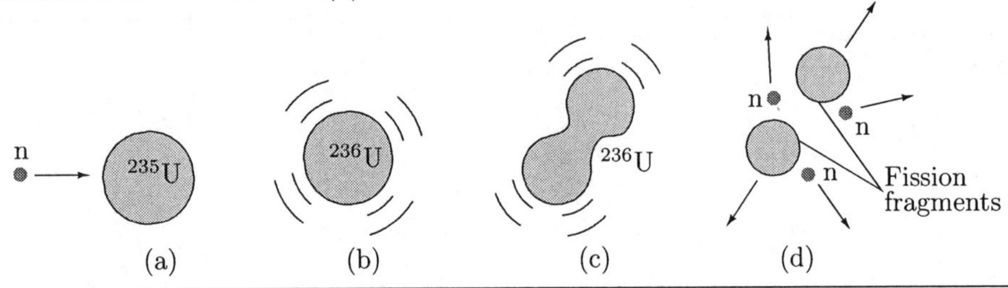

A typical fission reaction is $\text{n} + {}^{235}_{92}\text{U} \rightarrow ({}^{236}_{92}\text{U})^{\star} \rightarrow {}^{141}_{56}\text{Ba} + {}^{92}_{36}\text{Kr} + 3\text{n} + \text{KE}$. Typically, about 200 MeV of energy is released in each fission. The two medium-mass nuclei have roughly the same neutron-proton ratio (N/Z) as the original nucleus, so they lie above the stability line of the N versus Z plot. Consequently, these fission fragments undergo further decay, usually beta decay, until a stable N/Z ratio is reached.

The neutrons emitted in a fission reaction are available to induce more fissions in the sample. The average number of neutrons from each fission that succeed in producing a subsequent fission is called the **reproduction constant** k. If $k < 1$, the nuclear reactions die out. But if each of the two or three neutrons released in a fission results in a further fission—that is, $k > 1$—the number of fissions will increase exponentially. In this case, all the available energy is released in an uncontrolled manner during a very short time interval, resulting in a nuclear explosion.

A **nuclear fission reactor** is designed so that, on the average, one neutron from a fission reaction produces only one new fission, that is, $k = 1$. In this way the fission reactions are controlled, the reactor is self-sustaining, and the excess energy, which appears in the form of heat, is used to heat water to make steam to drive turbines that generate electrical power.

Because the fission of ${}^{235}_{92}\text{U}$ (for example) is induced most easily by low-energy neutrons, nuclear fission reactors employ **thermal neutrons**, which have energies $kT \approx 0.025$ eV. A thermal neutron results when the energy of a fission-produced neutron is reduced from its initial value of about 1 MeV to a thermal value by repeated collisions with light-weight nuclei in a **moderator**. The moderator, which may be water or carbon, is placed around the fissionable material in the core of the reactor.

In a **fusion reaction** two light nuclei ($A < 20$) fuse to form a heavier nucleus, with a release of energy. An example of a fusion reaction is the combination of deuterium and tritium to form a helium nucleus and a neutron: ${}^{2}\text{H}+{}^{3}\text{H} \rightarrow {}^{4}\text{He}+\text{n}+17.6$ MeV. The energy per unit mass released in a fusion reaction is larger than in a fission reaction. For example, in the deuterium-tritium fusion, (17.6 MeV)/(5 nucleons) = 3.52 MeV/nucleon is released. As a comparison, in a typical fission reaction, about 200 MeV is released per 200 nucleons, giving roughly 1 MeV per nucleon released.

For fusion to occur, the particles must be heated to a temperature corresponding roughly to a thermal energy $kT = 10$ keV, or about 10^8 K. At this temperature, atoms are separated into a gas of positive ions and negative electrons called a **plasma**. Temperatures of 10^8 K occur in stars, where fusion reactions play an important part in a star's evolution, and are the primary sources of a star's energy. On Earth, the confinement of a plasma for a period of time long enough for fusion to take place is a problem that we are only beginning to solve, so practical fusion reactors will probably not be available for several decades.

Important Derived Results

Typical fission reaction	$n + {}^{235}_{92}U \rightarrow \left({}^{236}_{92}U\right)^{\star} \rightarrow {}^{141}_{56}Ba + {}^{92}_{36}Kr + 3n + KE$
Deuterium-tritium fusion reaction	${}^{2}H + {}^{3}H \rightarrow {}^{4}He + n + 17.6\ MeV$

Common Pitfalls

> Do not confuse nuclear magnetic resonance with nuclear fission or nuclear fusion.

7. TRUE or FALSE: Fission is induced most readily with neutrons of high energy of the order of 1 MeV.

8. What is the difference between fission and fusion?

Try It Yourself #7

Assuming that in a fission reactor a neutron loses half its energy in each collision with an atom of the moderator, determine how many collisions are required to slow a 200-MeV neutron to an energy of 0.04 eV.

Picture: After each collision, $K_f = 0.5K_i$ and K_f becomes the initial kinetic energy for the next collision.

Solve:

Write an expression for the kinetic energy after n collisions.	
Find the ratio of the final and initial kinetic energies.	
Take the log of both sides of the expression in the previous step and solve for n.	$n \approx 33$

Check: This number is neither excessively large or small, so seems to make sense.

Try It Yourself #8

How many days will it take for 1.00 kg of $^{235}_{92}U$ to be used up in a nuclear reactor that generates 50.0 MW of power, if each fission produces 200 MeV of energy?

Picture: Use dimensional analysis to determine the number of fissions per second required to generate 50.0 MW. Use Avogadro's number to determine the number of uranium nuclei.

Solve:

Determine the number of fissions per second required to produce 50.0 MW of power.	
Determine the number of uranium atoms in the 1.00 kg of fuel.	
Use dimensional analysis with the previous two results to determine the time it takes to use up the fuel.	$t = 19$ days

Check: The units all work out properly.

Taking It Further: This is perhaps a small to medium-size power plant. Clearly it needs a regular supply of enriched uranium to continue operations. This is on reason enrichment programs are so important.

QUIZ

1. TRUE or FALSE: The strong nuclear force holds electrons inside a nucleus until the electrons are emitted in beta decay or nuclear reactions.
2. TRUE or FALSE: The cross section for a particular reaction depends only on the size of the nuclei being bombarded and not on the energy of the incident particles.
3. What is the difference between atomic number and mass number?
4. Explain how variations in the reproduction constant affect the operation of a nuclear reactor.
5. Why are thermal neutrons used in nuclear reactors?

6. The following fission reaction of ${}^{235}_{92}\mathrm{U}$ takes place: ${}^{235}_{92}\mathrm{U} + \mathrm{n} \rightarrow \left({}^{236}_{92}\mathrm{U}\right)^{\star} \rightarrow {}^{141}_{56}\mathrm{Ba} + {}^{92}_{36}\mathrm{Kr} + 3\mathrm{n}$. Find the Coulomb energy of the two fully ionized fission fragments, assuming that they are spheres just touching each other.

7. You want to produce fusion with deuterium nuclei. Estimate the temperature required to bring two deuterium nuclei together so that they can fuse. How much energy is released when they fuse?

Chapter 41

Elementary Particles and the Beginning of the Universe

41.1 Hadrons and Leptons

In a Nutshell

What we now call atoms were given their name from the ancient Greek word atomos, which means "indivisible." At the time, nothing was known about protons, neutrons, and electrons. Since the discovery of those particles in the late 1800s and early 1900s, we have developed an understanding that there are many more fundamental particles than even these. Furthermore, most of these particles consist of even more elementary particles called quarks. This chapter examines some of the physics of these myriad particles as we currently understand it. However, the story of elementary particles is by no means finished, so keep in mind that the particle picture presented here could change with new discoveries and new ways of thinking.

One way of classifying elementary particles is according to the way they participate in interactions between particles. The word "interaction" is used instead of "force" because the quantum picture of how elementary particles interact is quite different from the traditional action-at-a-distance forces of classical physics. Four basic interactions exist in nature:

1. The strong nuclear interaction
2. The electromagnetic interaction
3. The weak (nuclear) interaction
4. The gravitational interaction

The **strong nuclear interaction** is responsible for the attractive force that holds nucleons together in the nucleus. The interaction has a short range, its strength dropping to zero when participating particles are more than a few femtometers apart. The interaction is independent of the charge of the nucleons—that is, the proton-proton, neutron-proton, and neutron-neutron forces are all about equal. Particles that decay via the strong interaction have very short decay times of the order of 10^{-23} s which is roughly the time it takes for light to travel across a nucleus.

Hadrons are particles that interact via strong interactions. A hadron with a half-integral spin is called a **baryon**; neutrons and protons are common examples. A hadron with an integral or zero spin is called a **meson**; examples are the pion (π meson) and kaon (K meson). Baryons are the most massive of elementary particles, whereas mesons have masses intermediate between electrons and protons. The properties of some hadrons and their antiparticles are given in Table 41-1 on page 1391 of the text. An example of a strong interaction between hadrons is $\mathrm{p} + \pi^- \rightarrow \mathrm{n} + \pi^0$. As we shall discuss later in more detail, hadrons are thought to be constructed of entities called quarks.

The **weak interaction** occurs in beta decay (discussed in Chapter 40), which results in the production of an electron or positron, and a neutrino. The interaction is termed weak because its strength is about 10^{-13} times that of the strong interaction. The range of the weak interaction is about 100 times smaller than the short range of the strong interaction. Particles that decay via the weak interaction have lifetimes on the order of 10^{-10} s, which is much longer than the 10^{-23} s lifetimes of particles that decay via the strong interaction.

Leptons are particles that interact with each other via weak interactions. Currently leptons include the electron (e^-), the muon (μ^-), and the tau particle (τ^-), each of which has a charge equal to the charge on an electron. Each of these particles also has an associated neutrino: ν_e, ν_μ, and ν_τ, respectively. The neutrinos have a very small mass, currently thought to be less than 16 eV/c^2, and are electrically neutral. All these particles also have an antiparticle with the same mass and spin, but opposite charge. Leptons are all spin-$\frac{1}{2}$ particles. Unlike hadrons (which are composed of quarks), leptons are not composed of more basic particles and so can be regarded as truly elementary. An example of a weak interaction is the beta decay of a neutron (hadron) into a proton (hadron) plus two leptons: $\mathrm{n} \rightarrow \mathrm{p} + e^- + \bar{\nu}_e$.

Common Pitfalls

- Do not confuse hadrons with leptons. Understand which particles are hadrons and which are leptons.
- Do not confuse particles that decay via the strong interaction with those that decay via the weak interaction. The decay time for a strong interaction is of the order of 10^{-23} s, whereas the decay time for a weak interaction is of the order of 10^{-10} s.

1. TRUE or FALSE: Mesons and leptons are baryons.

2. What distinguishes a hadron from a lepton?

41.2 Spin and Antiparticles

In a Nutshell

Elementary particles are classified according to their spin. Spin-$\frac{1}{2}$ particles (or $\frac{3}{2}$, $\frac{5}{2}$, ...) are called **fermions**. They obey the Pauli exclusion principle—only one fermion can be in a given quantum state. Examples of fermions are electrons, protons, neutrons, and neutrinos. Particles with integer spin (0, 1, 2, ...) are called **bosons**. They do not obey the Pauli exclusion principle—any number of bosons can be in a given quantum state. Examples of bosons are mesons and photons.

Every particle has an **antiparticle** with the same mass and spin but opposite charge. A particle is usually an entity, such as an electron or proton, that is found naturally in our part of the universe. The antiparticle of a particle, such as a positron or an antiproton, is an entity that is not found in nature, and when formed quickly annihilates itself with its corresponding particle.

A single photon with energy equal to at least the rest energy of a particle and its corresponding antiparticle can create a particle–antiparticle pair. Similarly, when a particle and antiparticle annihilate, they produce two photons moving in opposite directions.

Common Pitfalls

- Do not confuse fermions with bosons. Fermions are spin $\frac{1}{2}$, $\frac{3}{2}$, $\frac{5}{2}$, ...particles and obey the Pauli exclusion principle. Bosons are spin 0, 1, 2, ...particles and do *not* obey the Pauli exclusion principle.

3. TRUE or FALSE: Particles and antiparticles have the same mass and charge.

4. A positron is stable, that is, it does not decay. Why, then, does a positron have only a short existence?

41.3 The Conservation Laws

In a Nutshell

In an elementary particle reaction or decay, the familiar conservation laws hold, in addition to some new conservation laws.

Conservation of mass-energy. The sum of the rest masses and kinetic energies of the particles before a reaction must be equal to the sum of the rest masses and kinetic energies of the particles after the reaction: $\sum\left(mc^2+K\right)_{\text{before}}=\sum\left(mc^2+K\right)_{\text{after}}$.

Conservation of linear momentum. The linear momentum $\vec{\boldsymbol{p}}$ of the particles before and after a reaction must be the same: $\sum(\vec{\boldsymbol{p}})_{\text{before}}=\sum(\vec{\boldsymbol{p}})_{\text{after}}$.

Conservation of angular momentum ($\vec{\text{spin}}$). Spin is a form of angular momentum. In any reaction, the vector sum of the spins of the particles after the reaction must equal the vector sum of the spins of the particles before the reaction: $\sum\left(\vec{\textbf{spin}}\right)_{\text{before}}=\sum\left(\vec{\textbf{spin}}\right)_{\text{after}}$.

Conservation of charge. The net charge Q before and after a reaction must be the same: $\sum(Q)_{\text{before}}=\sum(Q)_{\text{after}}$.

In addition, two conservation laws hold especially for baryons and leptons. Baryon numbers are assigned to particles as follows:

$$B=\begin{cases}+1 & \text{for all baryons}\\ -1 & \text{for all antibaryons}\\ 0 & \text{for all other particles}\end{cases}$$

Lepton numbers are assigned with the following rules:

$$L=\begin{cases}+1 & \text{for all leptons}\\ -1 & \text{for all antileptons}\\ 0 & \text{for all other particles}\end{cases}$$

This leads us to two more conservation laws that can be seen in the neutron decay:

$$\begin{aligned}\mathrm{n}&\rightarrow \mathrm{p}+\mathrm{e}^-+\bar{\nu}_e\\ B&:+1=+1+0+0\\ L&:0=0+1-1\end{aligned}$$

Conservation of baryon number. The net baryon number B before and after a reaction or decay must be the same: $\sum(B)_{\text{before}}=\sum(B)_{\text{after}}$.

Conservation of lepton number. The net lepton number L before and after a reaction or decay must be the same: $\sum(L)_{\text{before}}=\sum(L)_{\text{after}}$.

Another number that obeys a conservation law is **strangeness**. The notion of strangeness was introduced to account for unexpected reactions that involved baryons and mesons. For example, kaons and pions interact via strong nuclear reactions, so physicists expected that the decay $K^0 \rightarrow \pi^+ + \pi^-$ would proceed via the strong interaction with a decay time of the order of 10^{-23} s. Instead, the decay time is of the order of 10^{-10} s, a time that is characteristic of the weak interaction. Because of such experiments, a strangeness number S was assigned to the strange particles, as shown in Figure 41-3 on page 1399 of the text.

Conservation of strangeness. In a strong interaction the net strangeness number S before and after the reaction is the same. An example of this is the reaction

$$\begin{aligned} &\mathrm{p} + \pi^- \rightarrow \mathrm{K}^0 + \Lambda^0 \\ &S: 0 + 0 = +1 - 1 \end{aligned}$$

For weak interactions, a **strangeness selection rule** holds: in a weak interaction the strangeness number changes by zero or 1, that is: $\Delta S = 0, \pm 1$. An example of this is the weak decay of the Λ^0 particle formed in the previous reaction:

$$\begin{aligned} &\Lambda^0 \rightarrow \mathrm{p} + \pi^- \\ &S: -1 \neq 0 + 0 \end{aligned}$$

Fundamental Equations

Conservation of mass-energy	$\sum (mc^2 + L)_{\text{before}} = \sum (mc^2 + K)_{\text{after}}$
Conservation of linear momentum	$\sum (\vec{p})_{\text{before}} = \sum (\vec{p})_{\text{after}}$
Conservation of spin	$\sum (\vec{\text{spin}})_{\text{before}} = \sum (\vec{\text{spin}})_{\text{after}}$
Conservation of charge	$\sum (Q)_{\text{before}} = \sum (Q)_{\text{after}}$
Conservation of baryon number	$\sum (B)_{\text{before}} = \sum (B)_{\text{after}}$
Conservation of lepton number	$\sum (L)_{\text{before}} = \sum (L)_{\text{after}}$
Conservation of strangeness (strong interactions)	$\sum (S)_{\text{before}} = \sum (S)_{\text{after}}$
Strangeness selection rule (weak interactions)	$\Delta S = 0, \pm 1$

Common Pitfalls

> Watch out for the change of strangeness number in reactions involving strange particles. In strong interactions, strangeness is conserved ($\Delta S = 0$). In weak interactions, strangeness may be conserved ($\Delta S = 0$) or may change by one unit ($\Delta S = \pm 1$).

5. TRUE or FALSE: Lepton, baryon, and strangeness numbers are conserved in all reactions or decays.

Try It Yourself #1

Determine the unknown particle X in the strong reaction: $p + \pi^- \to K^0 + X$.

Picture: Apply the conservation laws to determine Q, s, B, L, and S for the unknown particle.

Solve:

Apply conservation of charge.	
Apply conservation of spin.	
Apply conservation of baryon number.	
Apply conservation of lepton number.	
Apply conservation of strangeness	
Compare these quantum numbers with the available particles to determine the possibilities.	$X = \Lambda^0$ or $X = \Sigma^0$

Check: These are the only particles that have the required $Q = 0$, $s = \frac{1}{2}$, $B = 1$, $L = 0$, and $S = -1$.

Taking It Further: How might you be able to distinguish between these Λ^0 and Σ^0 particles to determine which one was present for a given reaction?

Try It Yourself #2

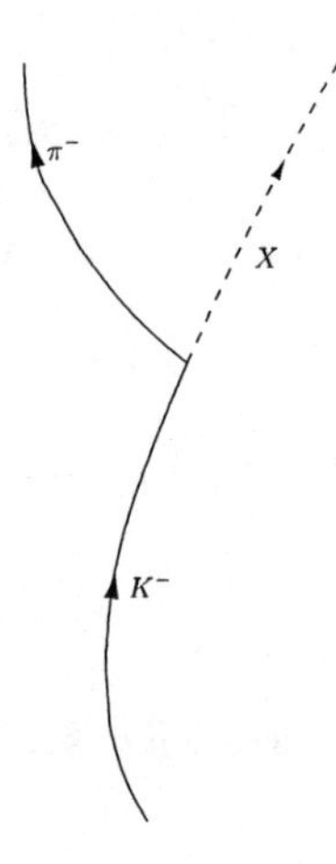

The solid lines in the figure show the tracks of two charged particles in a photograph of a reaction that occurred in a bubble chamber, where a magnetic field was directed into the paper. Determine the identity of the unknown neutral particle X, shown by the dashed line, that was one of the reaction products but was not observed in the bubble chamber.

Picture: Write an expression for the decay shown, and apply the conservation laws to determine the quantum numbers of the unknown particle.

Solve:

Write an expression for the decay shown.	
Apply conservation of charge.	
Apply conservation of spin.	
Apply conservation of baryon number.	
Apply conservation of lepton number.	
Apply the strangeness selection rule.	
Match the quantum numbers to the known particles.	

Use mass-energy conservation to narrow down the identity of the particle. Find the mass of the given particles, and the mass of the possible particles identified above.	
Determine which particle is allowed.	π_0

Check: This is the only particle that obeys all conservation rules and the strangeness selection rule.

Taking It Further: How can you tell that both the K and π particles are negatively charged?

41.4 Quarks

In a Nutshell

Hadrons (baryons and mesons) are thought to be made up of fundamental particles called **quarks**. It is convenient to divide our discussion of quarks into two parts: "old" quarks (the picture that developed before the mid-1960s) and "new" quarks (the understanding that has emerged since then).

The old quark picture contains three quarks with flavors labeled u, d, and s (for up, down, and strange) and their three antiquarks; each has a spin, charge (which is a fraction of the electron charge), baryon number, and strangeness number, as listed in Table 41-3 on page 1401 of the text. Hadrons are composed of quark combinations according to the following rules:

- Mesons consist of a quark–antiquark pair.
- Baryons consist of three quarks.
- Antibaryons consist of three antiquarks.

These simple quark-combination rules accounted for the structure of all the hadrons that were known before the mid-1960s.

To demonstrate these rules, consider the Ξ^- particle. It is currently believed that the quark structure for the Ξ^- particle is dss. By examining the properties of each of these quarks in Table 41-3 on page 1401 of the text, you should convince yourself that the quarks combine to give a net charge $Q = -e$, baryon number $B = 1$, strangeness number $S = -2$, and an overall spin of $\frac{1}{2}$ if the spins can be combined as vectors. These properties are exactly those of the Ξ^- particle.

By the late 1960s physicists realized that more than three quarks (and their antiquarks) were needed to explain new developments in the burgeoning elementary particle picture. So they added three new quarks, labeled c, t, and b (for charm, top, and bottom, or, what some people like better, charm, truth, and beauty). Each carries a new quantum number. The charmed quark has a **charm number** +1, the top quark has a **topness number** +1, and the bottom quark has a **bottomness number** +1, as listed in Table 41-3 on page 1401 of the text. Many more hadrons were discovered whose structure involved the new quarks, such as the charmed mesons $D^0(\bar{u}c)$, $D^+(\bar{d}c)$, $D^-(d\bar{c})$, and the charmed baryon $\Lambda_C^+(udc)$.

It is probably no accident that the picture of hadrons composed of six fundamental quarks meshes symmetrically with the picture of six fundamental leptons: the electron, muon, tau, and the three associated neutrinos. Deep-seated theoretical reasons, based on arguments of symmetry, suggest that there should be exactly six quarks and six leptons, though, of course, new developments could change this picture.

Do quarks really exist? That depends on what you mean by exist. The six-quark picture derives from elaborate theoretical arguments and explains the structure of the hundreds of hadrons that have been detected. But an isolated quark has never been observed, even though many experiments have looked for them. To account for this fact, a theoretical argument called **quark confinement** was developed. Quark confinement states that we will never observe isolated quarks because the force between quarks gets stronger as they separate, much like the force between balls at the ends of a rubber band. The enormous energy of this restoring force confines the quarks within the bounds of their hadron. Thus, the presently espoused quark picture follows from a theoretical underpinning which states that the particle on which the whole structure of the physical universe is based—the quark—can never be observed.

Common Pitfalls

- It can be very easy to confuse S (strangeness number), s (spin), and s (the strange particle). Make sure you are aware of the appropriate context.

6. TRUE or FALSE: Quarks have fractional multiples of the electronic charge.

7. How many quarks are there? Name them.

Try It Yourself #3

Determine the particle that is composed of the quark structure *uds*.

Picture: Look up the quantum numbers resulting from these quarks, and compare that to the quantum numbers of known particles. Remember that a combination of three quarks means the resulting particle is a baryon or antibaryon.

Solve:

Determine the charge quantum number that results from combining the three quarks.	

Determine the strangeness quantum number that results from combining the three quarks.	
Determine the spin quantum number that results from combining the three quarks.	
Compare these quantum numbers to the properties of known baryon and antibaryon particles to determine the possibilities.	Λ^0 or Σ^0

Check: The quantum numbers all work out.

Taking It Further: How would your answer change if the quark combination were instead *uss*? Explain.

Try It Yourself #4

What is the quark structure of a π^- meson?

Picture: Recall that a meson is composed of a quark–antiquark pair. Look up the quantum numbers of the meson and determine the quark–antiquark pair than can produce that combination.

Solve:

Determine which quark–antiquark pairs can combine to give the net charge of the π^- meson.	
Determine which quark–antiquark pairs can combine to give the net strangeness of the π^- meson.	

Determine which quark–antiquark pairs can combine to give the net spin of the π^- meson.	
Which quark–antiquark pair satisfies all three conditions above?	$\bar{u}d$

Check: The quantum numbers all work out properly.

41.5 Field Particles

In a Nutshell

The theoretical picture of the mechanism that produces the interaction between elementary particles—leptons or quarks—is quite different from the classical picture of forces. The elementary particle theory states that interactions between particles take place via the exchange of virtual **field particles** or **field quanta**. The word "virtual" is used because the field quanta are not observed directly but are inferred from theoretical considerations. One particle emits a field quantum that is absorbed by another particle. The second particle then sends the field quantum back to the first particle. The force between the two particles results from this continuous game of catch with the field quantum, which is said to **mediate** the interaction. An analogy can be seen in a game of catch between two persons on frictionless ice. When the two persons throw a ball back and forth, each experiences a repulsive force, as in Figure (a). When the two persons skillfully throw a boomerang back and forth, each experiences an attractive force, as in Figure (b).

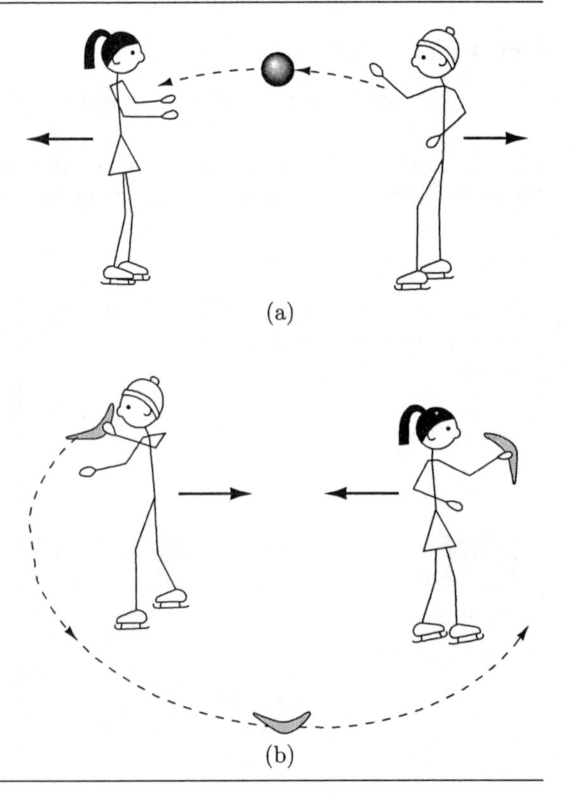

The four fundamental interactions are mediated by four different field quanta.

1. **Electromagnetic interaction**. The field quantum is the photon, which has rest mass 0, charge 0, and spin 1.
2. **Gravitational interaction**. The field quantum is the graviton, which has rest mass 0, charge 0, and spin 2. The graviton has not been observed experimentally.
3. **Weak interaction**. There are three field quanta called **vector bosons**: W^+, W^-, and Z^0. The W^+ and its antiparticle W^- have identical masses of 80.22 GeV/c^2, charges of $+e$ and $-e$ respectively, and spin 1. The Z^0 has a mass of 91.19 GeV/c^2, charge 0, and spin 1.
4. **Strong interaction**. The field quantum is the gluon, which has rest mass 0, charge 0, and spin 1. Gluons are responsible for the strong force between quarks. A gluon has not been observed experimentally.

Note that all the field quanta have integer spins, so all are bosons.

The development of the theory of strong interactions has resulted in the introduction of another new quantum number called **color charge**. Quarks are said to come in three colors—red, blue, and green—which are the "charges" responsible for quark interactions. The field theory that describes these colors is called **quantum chromodynamics (QCD)**.

Common Pitfalls

> Do not confuse field quanta with the fundamental lepton or quark particles. Field quanta are particles that are exchanged between leptons or quarks, thereby producing the force exerted between two fundamental particles.

8. TRUE or FALSE: Gluons are field quanta that "glue" quarks together, that is, gluons are responsible for the strong force between quarks.

9. Describe how field quanta produce forces between particles.

41.6 The Electroweak Theory

In a Nutshell

The **electroweak theory** is an attempt to unite the electromagnetic and weak interactions into a single more fundamental interaction. At very high particle energies, $\gg 100$ GeV, the single electroweak interaction is mediated by four field quanta: the W^+ and W^- and new quanta called B^0 and W^0, which cannot be observed directly. The observed Z^0 and the photon are formed from combinations of the B^0 and W^0 bosons. The united electromagnetic and weak interaction have equal strength and a range of less than 10^{-19} m. As the particle energy decreases, the equal strength symmetry between the interactions is broken, and the single electroweak interaction becomes two separate interactions, electromagnetic and weak.

Common Pitfalls

10. TRUE or FALSE: According to the electroweak theory, at very high energies symmetry considerations show that the electromagnetic and weak interactions are part of a single electroweak interaction.

11. What does the electroweak theory do?

41.7 The Standard Model

In a Nutshell

The particle picture shows that the fundamental building blocks of all matter are leptons and quarks. Forces between these fundamental particles result from the exchange of field quanta in one of four basic interactions—strong, electromagnetic, weak, or gravitational—with the weak and electromagnetic interactions being combined at very high energies into the single electroweak interaction. Besides the familiar electric charge of electromagnetic interactions and the mass (charge) of gravitational interactions, there are weak flavor charges carried by leptons and color charges carried by quarks and gluons. All of this—and much more that we have not mentioned—is the **standard model**.

The electroweak theory joins the electromagnetic and weak interactions into a single electroweak interaction. Pushing this idea still further, many people have been trying to unite the electroweak and the strong interactions under a single **grand unification theory** (**GUT**). Thus far no one has been able to make a GUT work.

Common Pitfalls

> Do not confuse the electroweak theory with grand unification theories (GUTs). The electroweak theory unifies the electromagnetic and weak interactions and is generally thought to be established. Grand unification theories are attempts to produce still further unification by combining electromagnetic, weak, and strong interactions into a single theory. Thus far no one has developed a successful GUT.

12. TRUE or FALSE: A bumper sticker reading "Particle physicists have GUTs!" means that particle physicists have an enormous intestinal fortitude for trying to understand the complex world of elementary particles.

13. What is the hope for GUTs?

41.8 The Evolution of the Universe

In a Nutshell

One of the properties of a galaxy that can be readily measured is the redshift of its emitted radiation. If the redshift is associated with the Doppler effect alone, the recessional velocity of the galaxy from Earth can then be calculated. In addition, the distance of the galaxy can be determined from the apparent brightness of objects of known intrinsic brightness or luminosity. For example, Edwin Hubble used the known luminosity of Cepheid variable stars as a "yardstick" to gauge the distances to nearby galaxies that contain Cepheids. In many cases these distance yardsticks are not accurately known, resulting in uncertainties in the values of distances ascribed to the other galaxies. Hubble found that the velocity v of a receding galaxy (as determined from its redshift) is linearly related to the distance r of the galaxy from Earth (as determined from its brightness and galactic yardsticks) by **Hubble's law**: $v = Hr$, where H is the **Hubble constant**, which has units of reciprocal time. Hubble's law is illustrated here for a group of spiral galaxies.

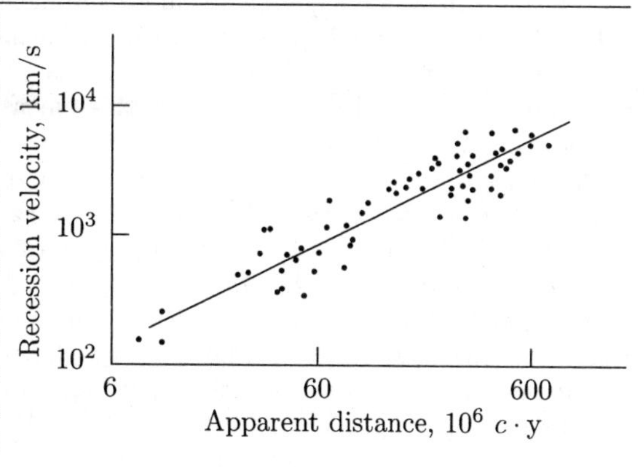

Because it is very difficult to measure astronomical distances, distances to only a small fraction of the galaxies in the observable universe have been determined. Moreover, the astronomical yardsticks are constantly being recalibrated as more data become accessible to astronomers. Currently, the value of the Hubble constant has been estimated to be $H = (22\text{ km/s})/(10^6\ c \cdot \text{y})$.

We have talked about Hubble's law as relating velocities and distances of galaxies measured from Earth, or equivalently from our Galaxy (the Milky Way). However, our Galaxy does not occupy a special place in the universe. Measurements made from any other galaxy should yield the same Hubble law with the same Hubble constant. Hubble's law has an enormously profound consequence. If you are located on any galaxy in the universe, all other galaxies are receding from you with a speed proportional to the distance from your galaxy. The numerical value of recessional speed of any galaxy from your galaxy is given by the present value of the Hubble constant. In short, the entire universe is expanding.

The presently observed abundance of helium in stars cannot be explained by nucleosynthesis in the stars. The helium abundance can be accounted for by fusion occurring in the extremely high temperatures in the early stages of the **Big Bang**. If the universe were expanding as a blackbody after the Big Bang, the **blackbody radiation** still remaining should have a temperature of around 3 K. Blackbody radiation of this temperature was experimentally detected in 1965 by Arno Penzias and Robert Wilson at the Bell Telephone Laboratories in New Jersey (for which they shared a Nobel Prize). Since then, more precise measurements have fixed the background blackbody radiation as having a temperature of 2.7281 K.

Out of many cosmological models that have been proposed to describe the structure and evolution of the universe, the **Big Bang** is the model that at present has obtained widespread acceptance. According to this model, our universe was born approximately 13.7 billion years ago in a giant explosion—the Big Bang—that occurred throughout all space, and has been expanding ever since. The Big Bang picture is supported by several experimental observations, including Hubble's law. There are more radio galaxies at far distances than close by. Since observations of farther distances correspond to earlier times, this means that there were more radio galaxies at earlier times than there are now, showing that the universe is indeed evolving and is not static.

The times discussed in the "very early" expanding universe are amazingly short, and the corresponding temperatures and energies are amazingly high. Before 10^{-43} s after the Big Bang, the four forces presently known today—gravitational, strong, electromagnetic, and weak—were unified in a single interaction described by a single (as yet unknown) theory. After 10^{-43} s, when the expanding universe had cooled to a temperature of about 10^{32} K, gravity broke free from this unification, and the remaining strong, electromagnetic, and weak forces remained unified in a single interaction described by grand unification theories (GUTs). At this stage, elementary particles had energies of the order of 10^{19} GeV. As the still-expanding universe cooled further to 10^{27} K at around 10^{-35} s, the strong force split away from the GUTs groups, leaving the electromagnetic and weak interactions still united as an electroweak force. Still later, around 10^{-10} s, the electromagnetic and weak forces parted company from their electroweak union, forming separate electromagnetic and weak forces, resulting in the four forces as we know them today.

As the universe kept expanding and cooling, the predominance of different elementary particles changed from one epoch to another. Quarks and leptons that were initially indistinguishable from each other in a "quark soup" became separate entities. "Ordinary" particles such as protons, neutrons, and electrons eventually emerged and combined into atoms and molecules. Eventually, as stars and galaxies formed, the density of matter grew larger than the density of radiation, resulting in the matter-dominated universe as we know it now, about 10-20 billion years after the Big Bang.

Physical Quantities and Their Units

Hubble constant	$H = \dfrac{22 \text{ km/s}}{10^6\ c \cdot \text{y}}$
Temperature of the present background blackbody radiation	$T = 2.7281$ K

Important Derived Results

Hubble's law	$v = Hr$

Common Pitfalls

14. TRUE or FALSE: The Big Bang refers to the enormous explosion of a supernova.

15. How is the presently observed background blackbody radiation related to the theory of the Big Bang?

QUIZ

1. TRUE or FALSE: A 1.00-MeV photon cannot create an electron–positron pair.
2. TRUE or FALSE: A quark never has more than one nonzero strangeness, charm, topness, or bottomness number.
3. When a positron and electron annihilate at rest, why must more than one photon be created?
4. How is the presently observed abundance of helium in stars related to the theory of the Big Bang?
5. Hadrons are built from quarks. An isolated quark has never been observed. How does particle theory explain this?

6. What is the quark structure of a K^0 meson?

7. Which of the following two possibilities for the weak decay of a Σ^- particle are possible? Why?

   (a) $\Sigma^- \rightarrow \pi^- + p$

   (b) $\Sigma^- \rightarrow \pi^- + n$

Answers to Problems

Chapter 34

COMMON PITFALLS

1. True
2. A threshold photon energy $E_t = hf_t = hc/\lambda_t$ equal to the work function ϕ is required to eject the least tightly bound electrons from the surface of the emitting material C: $hc/\lambda_t = \phi$. A photon with less energy than this does not eject any electrons.
3. False. Because the de Broglie wavelength $\lambda = h/p$, particles with the same momentum have the same wavelength.
4. The velocity of the electron is related to its kinetic energy by $K = \frac{1}{2}mv^2$ or $v = \sqrt{2K/m}$. The de Broglie wavelength is $\lambda = h/(mv) = h/\sqrt{2mK}$.
5. False. The probability is proportional to the square of the wave function.
6. The size dx of the spatial interval and the probability density $P(x)$ define the probability. The probability of finding an object in an interval dx is given by $P(x)\,dx$.
7. False. The more uncertain a particle's momentum, the more *precisely* we may be able to locate its position.
8. Such an experiment does not exist. According to Bohr's principle of complementarity, which follows from the uncertainty principle, you cannot measure both the wave and the particle aspects of an object in any single experiment.
9. True
10. As the length of the box increases, the ground-state energy for the particle is decreased and the separation of the energy levels is also decreased. If the length of the box is extended to infinity, the ground-state energy becomes zero, and there is a continuum of allowed energy states.
11. True
12. (a) $E_n \propto n^2$; (b) $E_n \propto n$; (c) $E_n \propto -1/n^2$

TRY IT YOURSELF–TAKING IT FURTHER

1. Because the work function is inversely related to the threshold wavelength, a material with a higher work function would have a smaller threshold wavelength.
2. As the Compton angle is increased, the change in wavelength also increases, so at a scattering angle of 180°, the scattered wavelength is the largest possible. This corresponds to the smallest possible energy of the scattered wavelength.
3. $K = 73.6$ eV. We should expect this kinetic energy to be larger than that of the neutron because the mass of the electron is much smaller.
5. The wave function is normalized. You can determine this by evaluating $\int_{-\infty}^{\infty} \psi^2(x)\,dx$, which equals 1.
6. The answer does not change at all. Both the $n = 2$ and $n = 4$ probability functions contain an even number of antinodes. As a result, $x = 0.25L$ is the position at which the 25 percent probability occurs.
7. Electrons with energy less than that of the ground state are *not* trapped in the potential well. Quantization tells us that only particular energies can be trapped, and in particular, there is a minimum energy that particles must have to remain contained in the box. This is part of the reason that the classical "electron clouds" around atoms are so much larger than the nuclei.
8. The expectation value of x is 0. This is because the resulting integral is an odd function evaluated from $-\infty$ to ∞. You should have anticipated that the expectation value of x is zero, because that is the average position value of a harmonic oscillator.
9. No. The energy levels of the hydrogen are not equally spaced.

QUIZ

1. True
2. True
3. The velocity of the electron is related to its kinetic energy by $\frac{1}{2}mv^2 = K$ or $v = \sqrt{2K/m}$. The de Broglie wavelength is $\lambda = h/(mv) = h/\sqrt{2mK}$.
4. The larger the work function ϕ of the emitting surface C, the more energy is required to release electrons from the emitter's surface. Hence, for incident photons of a given frequency, and therefore a given energy, the larger the work function ϕ, the lower the maximum kinetic energy of the emitted electrons. This is seen from the energy balance in Einstein's photoelectric equation $hf = K_{\text{max}} + \phi$.
5. Despite its name, the expectation value of x does not provide the expected location of a particle. It is better understood as the the average value of x. Since the probability density of the first excited

state is an even and symmetric function about the center of the box, it is equally probable that the particle spend time on either side of center. So the average position of the particle is the center, even though the particle will never be found there.

6. 247 nm
7. $\langle x \rangle = L/2$

Chapter 35

COMMON PITFALLS

1. False. For a given potential energy function, there is a family of wave functions ψ_n that are solutions to the Schrödinger equation.
2. The wave functions for a particle in a box are sinusoidal in nature. The wave function must equal zero at both ends of the box. This gives rise to a discrete set of allowable wavelengths, frequencies, and energies for particles in a box.
3. False. A quantum-mechanical analysis predicts tunneling into this classically forbidden region.
4. There are an infinite number of quantized energy levels in an infinite square well, but there are a finite number of quantized energy levels in a finite square well.
5. False. The ground-state wave function varies like an exponential Gaussian distribution.
6. The energy levels of a particle in a harmonic energy potential are evenly spaced, with a spacing of $\hbar\omega_0$.
7. False. On the right side the wave function follows an exponential decay with no associated wavelength.
8. The amplitude is greater on the side where the particle is incident. As the particle penetrates the barrier, the amplitude of its wave function decreases exponentially, so it is smaller when it emerges from the barrier.
9. False. Degeneracy refers to a situation in which a given energy level can be associated with two or more wave functions.
10. False. Fermions and bosons are two different classes of particles. A fermion is described by an antisymmetric wave function, and no two fermions can have the same quantum numbers. Bosons are described by symmetric wave functions, and it is possible for two bosons to have the same quantum numbers.
11. If two particles have the same quantum numbers, the antisymmetric wave function is identically zero. This is not the case for a symmetric wave function.

TRY IT YOURSELF–TAKING IT FURTHER

3. The only ways to have a zero percent probability that the electron penetrates the barrier are to let $U_0 = \infty$, or $a = \infty$. Neither of these is very practical. However, it is not too difficult to make the probability of penetration small enough to ignore—just how small that is depends on the situation.

QUIZ

1. False. A symmetric function is not zero when the quantum numbers of the two particles are the same.
2. True
3. In classical physics, the kinetic energy of a particle with total energy E is given by $K = E - U(x)$, where $U(x)$ is the potential energy of the particle. The kinetic energy $K = \frac{1}{2}mv^2$ must always be positive, so it is impossible to have $E < U(x)$. As an example to visualize this, think of a ball having a total energy E rolling up the sloping side of a hill described by a potential energy function $U(x)$. There will be a value x_0 where $U(x_0) = U_0 = E$. When the ball reaches the point $E = U_0$, where its total and potential energies are equal, $K = E - U_0$ becomes zero and the ball stops moving upward and rolls back. The ball can never enter the forbidden region where $U(x) > U_0 = E$.
4. The wave function for two fermions is antisymmetric, while the wave function for bosons is symmetric. This means that fermions satisfy the Pauli exclusion principle, while bosons do not satisfy the Pauli exclusion principle.
5. The wavelength is the same on both sides of the barrier, as long as the barrier is symmetric.
6. $E = 11.5$ MeV
7. $E_{11} = \left[\hbar^2\pi^2/(2m)\right](10/9L^2)$; $E_{12} = \left[\hbar^2\pi^2/(2m)\right](13/9L^2)$; $E_{13} = \left[\hbar^2\pi^2/(2m)\right](2L^2)$. Each of these states has a degeneracy of 1.

Chapter 36

COMMON PITFALLS

1. True

2. From the expression $E_n = -Z^2 E_0/n^2$ the energies are seen to vary as $1/n^2$—that is, as n increases the energy becomes less and less negative, until it reaches zero at $n = \infty$.
3. False. The relationship is $L = \sqrt{\ell(\ell+1)}\hbar$.
4. False. The hydrogen wave functions $\psi_{2\ell m}$ are spherically symmetric only for $\ell = 0$ and $m_\ell = 0$.
5. $\psi^2(r)\,dV = \psi^2(r)r^2 \sin\theta\,dr\,d\theta\,d\phi$ is the probability of finding an electron in the volume element $dV = r^2 \sin\theta\,dr\,d\theta\,d\phi$. You get a factor of 4π when the angular integration is performed, leaving you with $\psi^2(r)r^2 4\pi\,dr = P(r)\,dr$ as the probability for finding an electron between r and dr. The volume of the corresponding shell is $4\pi r^2\,dr$.
6. True
7. The interaction between the spin and orbital magnetic moments splits each energy level with $\ell > 0$ into closely spaced energy levels. For a single electron, for example, there is a doublet splitting corresponding to $j = \ell + \frac{1}{2}$ and $j = \ell - \frac{1}{2}$. Transitions from one pair of closely spaced levels to another pair of closely spaced levels result in the observed fine structure.
8. False. Elements with filled shells are chemically inactive and difficult to ionize.
9. True
10. The wavelengths of the characteristic peaks are unaffected by increasing the accelerating voltage of the bombarding electrons because the characteristic spectrum depends only on the electron energy levels in the atoms of the target material.

TRY IT YOURSELF–TAKING IT FURTHER

1. $\lambda_7 = 396.7$ nm
4.

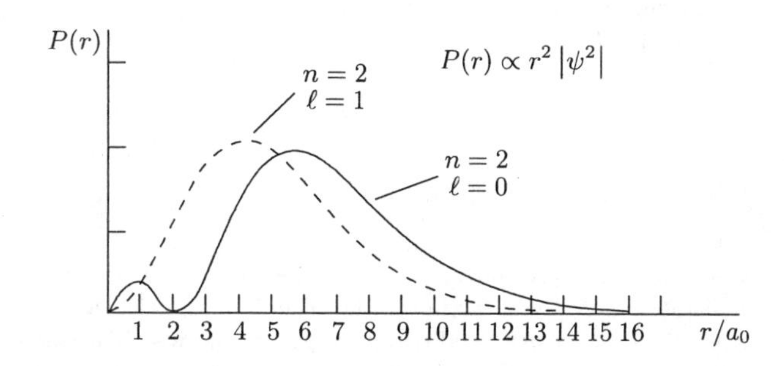

QUIZ

1. False. The wavelengths of spectral series are determined by fixing the value of n of the final energy level.
2. False. There can be two electrons with the given value of m_ℓ, one with $m_s = +\frac{1}{2}$ and one with $m_s = -\frac{1}{2}$.
3. From the expression $E_n = -Z^2 E_0/n^2$ the energies are seen to vary as Z^2, where Z is the number of protons in the nucleus.
4. The more energetic the bombarding electrons, the larger the maximum energy $E_{max} = hf_{max} = hc/\lambda_m$ of the photons emitted in the bremsstrahlung process. This means that the cutoff wavelength λ_m decreases as the accelerating voltage of the bombarding electrons is increased, resulting in more and more characteristic wavelengths being observed.
5. When a K-shell ($n = 1$) electron is knocked out of an atom, the outer electrons that can make transitions to the empty K shell see a nucleus with charge $+Ze$ and a charge $-e$ on the remaining electron in the K shell, or a net charge of $(Z-1)e$. The Moseley relation arises when the charge $(Z-1)e$ is used in the Bohr theory.
6. 0.272
7. $B = 0.051$ T

Chapter 37

COMMON PITFALLS

1. False. There is usually a mixture of ionic and covalent bonding.
2. Exclusion-principle repulsion keeps the ions separated.
3. False. The energy of the lowest vibrational energy level is $E_0 = \frac{1}{2}hf$, the zero-point energy.
4. Measure the energies of the rotational energy levels from the rotational spectrum. Then use $E = \ell(\ell+1)\hbar^2/(2I)$ to find the moment of inertia I.

TRY IT YOURSELF–TAKING IT FURTHER

1. The dipole moment should be larger. MgO is created with an ionic bond resulting from the sharing of two electrons, so the charge of each molecule is twice that for LiF. The separation of the two atoms will not be half that of LiF.
2. This charge is much less than $-2e$ because the two electrons of an H_2 atom are not located exactly between the hydrogen atoms.

QUIZ

1. False. It's the other way around.
2. False. Bands are seen in transitions between two vibrational energy levels because the transitions actually occur between closely spaced rotational energy levels that are built on each vibrational energy level. These transitions between the rotational energy levels result in closely spaced spectral lines that can be seen upon higher resolution.
3. The ionically bonded molecule has the larger dipole moment. In a molecule that has a covalent bond, the dipole moments, formed from each positively charged nucleus and from the negatively charged electrons between the nuclei, point in opposite directions and tend to cancel, resulting in a small overall dipole moment for the atom.
4. From the quantized vibrational energies, $E_\nu = \left(\nu + \frac{1}{2}\right) hf$, for $\nu = 0, 1, 2, \ldots$, the lowest energy corresponding to $\nu = 0$ has the nonzero value $E = hf/2$. Thus in its lowest energy state a molecule is not at rest but is vibrating with this zero-point energy about its equilibrium position.
5. The van der Waals bond arises from the attraction of molecular dipoles for each other. The molecular dipoles can be either permanent dipoles or induced dipoles.
6. $D = (ke^2/r_0) - R + A - I$
7. 0.227 eV

Chapter 38

COMMON PITFALLS

1. True
2. The potential energy of a given ion in a crystal due to another ion in the crystal is $U = \pm ke^2/R$, where R is the distance from the given ion to the other ion, and the sign depends on the charges of the two ions. Because the ions are arranged regularly, groups of ions of the same charge are located at the same distance from the given ion; so all ions in a particular group contribute the same value toward the overall potential energy. When the contributions from all the groups throughout the crystal are added to get the net potential energy of the given ion, the result is $U_{\text{net}} = -\alpha ke^2/r$, where r is the separation distance between neighboring atoms, and α is the Madelung constant, which depends only on the geometry of the crystal structure.
3. True
4. Mean velocity is the velocity of electrons occurring because of thermal motion. Drift velocity is the velocity that electrons acquire in addition to the mean velocity when an electric field is established in a conductor. The drift velocity is of the order of 10^{-9} times smaller than the mean velocity.
5. False. In the ground state, the N bosons all occupy the lowest energy level. Because of the exclusion principle, however, only two fermions can occupy the lowest energy level, and the other fermions occupy higher energy levels. Thus in the ground state the energy of N fermions is larger than the energy of N bosons.
6. The classical model assumes that electrons behave like Newtonian point particles obeying Maxwell–Boltzmann statistics. However, electrons have wave properties and obey the Pauli exclusion principle and quantum-mechanical Fermi–Dirac statistics.
7. False. It is called the band theory because the energy levels of many closely packed atoms split into groups of nearly continuous closely spaced energy levels called bands.
8. In an insulator, the energy gap between the filled valence band and the conduction band is so large that at normal temperatures few electrons are excited into the conduction band to participate in electrical conduction. In a semiconductor, the energy gap between the filled valence band and the conduction band is very small, so at normal temperatures even thermal energy can excite a large number of electrons into the conduction band to participate in electrical conduction, leaving holes in the valence band that also can participate in electrical conduction.
9. True
10. To obtain a forward-biased *pn*-junction diode, the positive terminal of a voltage source is connected to the *p* side of the diode, resulting in a large current. To obtain a reverse-biased diode, the positive terminal of a voltage source is connected to the *n* side of the diode, resulting in essentially no current.

11. False. A Cooper pair behaves like a spin-0 boson and does not satisfy the Pauli exclusion principle.
12. It is the temperature at which a superconducting material changes from its normal-conducting state to a superconducting state, in which it has zero resistance.
13. True
14. The Fermi–Dirac distribution function varies with temperature in the same manner as the Fermi factor: $f(E) = \left(e^{(E-E_F)/(kT)} + 1\right)^{-1}$.

TRY IT YOURSELF–TAKING IT FURTHER

2. $\rho = m_e v_{av}/(n_e e^2 \lambda) \Longrightarrow v_{av} = 1.64 \times 10^4$ m/s, about 10^9 times larger than the drift speed.

QUIZ

1. False. The number of electrons $n(E)\,dE$ with energies between E and $E + dE$ equals the number of states $g(E)\,dE$ between E and $E + dE$ multiplied by the Fermi factor $f(E)$ which gives the probability that a state is occupied: $n(E)\,dE = g(E)\,dE\,f(E)$.
2. False. The controlled addition of impurities to semiconductors is the predominant way of making semiconductor devices. But one should get rid of any other, unwanted impurities.
3. In a conductor, there is no energy gap between the valence and conduction bands, so the electrons in the conduction band can easily acquire enough kinetic energy to participate in electrical conduction. In an insulator, there is a large energy gap between the filled valence band and the conduction band, so almost no electrons are in the conduction band to participate in electrical conduction.
4. The Fermi factor $f(E)$ is a function of energy that gives the probability that a state of energy E is occupied.
5. When there is no voltage across a Josephson junction (formed by two superconductors), a dc current is observed; this is the dc Josephson effect. When a dc voltage is applied across a Josephson junction, an ac current is observed; this is the ac Josephson effect.
6. $N = 4.23 \times 10^{23}$
7. $E_F = 8.88$ eV. The material is magnesium.

Chapter 39

COMMON PITFALLS

1. True
2. No. The Michelson-Morley experiment was designed to measure the speed of Earth relative to the ether, a measurement that did involve high precision.
3. True
4. The pilot measures the time interval with a single clock that records the proper time between the beginning and end points. This proper time interval will be smaller than your time interval Δt.
5. False. Since both events occur at the same place, if one observer finds the two events to be simultaneous, then all other observers will also find them to be simultaneous. Disagreements about simultaneity arise only concerning events that are spatially separated in at least one of the reference frames.
6. Since the red and blue flashes are both emitted at the same place as determined in reference frame S', the time interval of 5.00 s determined by a single clock in S' is the proper time interval between the two events.
7. False. The relativistic velocity transformation gives the velocity of A relative to B to be $0.75c$.
8. The time dilation effect depends on the square of the speed, so the direction of the motion makes no difference. Since A and B move with the same speed over the same distance, they will be the same age when the triplets get together, and they will be younger than C, who stayed at home.
9. True
10. True
11. When the kinetic energy $K \ll mc^2$, you can use the expression $p = mu$ to a good approximation because in this limit relativistic speeds are not involved.

TRY IT YOURSELF–TAKING IT FURTHER

1. The negative time difference in the S' reference frame means that in that reference frame lightning bolt B appeared to strike *before* lightning bolt A. This completely counterintuitive but correct result can be arrived at only by applying relativistic transformations. By re-applying the transformation for t' and requiring that $\Delta t' = 0$, you find that an observer in reference frame S' will observe the events to occur at the same time if S' moves with a speed of $0.200c$ relative to frame S.
2. $L = 1.44 \times 10^5$ m. This distance is shorter than the track length measured by the track observer. It must be, because the car's speed is the same in both frames, but since less time elapses in the driver's frame of reference, the distance traveled as measured by the driver must be shorter.

4. 5.19 y. Simply divide the given distance (as measured on Earth) by the speed of the spaceship.
6. Consider the previous problem. If the neutron is "Earth" then the laboratory goes by at a speed of $0.8c$, which corresponds to rocket B in this problem. The neutron emits an electron in the forward direction, at a speed of $0.6c$, and this electron corresponds to rocket A in this problem. In the previous problem we determined the speed of the electron relative to the lab frame, which is then equivalent to the speed of rocket A relative to rocket B in this problem. Since two objects must always have the same speed relative to each other, our calculation of the speed of rocket B relative to rocket A better give the same answer.
7. $v \approx 46c$, which is greater than the speed of light, so clearly incorrect.
8. If the particle were a proton, then the full relativistic expression is required because the kinetic energy is on the same order as the rest energy of a proton.

QUIZ

1. True
2. True
3. If the object is moving relative to you, you measure its length by finding the difference between the coordinates of its end points at the same time.
4. Yes. Subtraction of the inverse Lorentz transformation expression for time gives $t'_A - t'_B = \gamma\left[(t_a - t_B) - (v/c^2)(x_A - x_B)\right]$. Suppose in reference frame S event A occurs after event B so that $t_A - t_B$ is positive. The time ordering $t'_A - t'_B$ between the two events as determined in reference frame S' can be positive, zero, or negative depending on the spatial separation $x_A - x_B$ in reference frame S and the relative velocity v.
5. When the total energy $E \gg mc^2$ or equivalently when the kinetic energy $K \gg mc^2$, you can use $p = E/c$ to a good approximation.
6. 125 MeV
7. 3.26 m

Chapter 40

COMMON PITFALLS

1. True
2. Isotopes are nuclides with the same atomic number Z but a different number of neutrons N, and so they have different mass numbers $A = Z + N$.
3. True
4. Consider a typical β decay, such as the decay of a neutron: $\mathrm{n} \rightarrow \mathrm{p} + \beta^- + \bar{\nu}_e$. If the electron ($\beta^-$) were the only decay product, application of conservation of energy and momentum to the two-body decay would require that the β^- particle be ejected with a single unique energy. Instead, it is observed experimentally that β^- particles are produced with energies that range from zero to a maximum value. Further, because the original neutron had spin $\frac{1}{2}$ conservation of angular momentum would be violated if the final decay products consisted of only the two particles p and β^-, each with spin $\frac{1}{2}$. To preserve conservation of energy and momentum, Pauli in 1930 postulated the existence of a third particle in the decay process, the neutrino, which was experimentally observed by Cowan and Reines in 1956.
5. True
6. The larger the nucleus of the bombarded element, the larger the cross section. Also, generally speaking the lower the speed of the incoming particle the larger the cross section. If the energy of the incoming particle happens to be near a resonance of the bombarded nucleus, then the cross section can be extremely large.
7. False. Fission occurs most readily with low-energy thermal neutrons (0.025 eV).
8. In fission, a heavy nucleus splits, usually by absorbing a neutron, into two lighter nuclei, with a corresponding release of neutrons and energy. In fusion, two light nuclei fuse to form a heavier nucleus, with a corresponding release of energy.

TRY IT YOURSELF–TAKING IT FURTHER

1. The materials we interact with are not made solely of nuclei; rather, they are made of atoms and molecules, which in addition to nuclei also have electrons. The orbital radii of the electrons is many orders of magnitude larger than the nuclear radius, and the mass of the electrons is quite small compared to that of a nucleon. The result is that atoms and molecules are much less dense than the nuclei.
2. If the binding energy were negative, that would mean that the nucleus gives up energy if it loses one or more nucleons. That means that the nucleus is not energetically stable, and actually prefers to give

up those nucleons. This process is called radioactive decay, and is the topic of the next section.

3. The radiation emission rate of radium is slightly less than 1 Ci.
5. Because the Q value is positive, this is an exothermic reaction, and energy is released by this reaction.

QUIZ

1. False. There are no electrons inside a nucleus. The strong nuclear force holds together the protons and neutrons inside the nucleus; it does not affect electrons.
2. False. The cross section for a particular reaction is a function of the energy of the bombarding particles and can be much less than the geometrical cross section of the target nuclei.
3. Atomic number Z equals the number of protons in a nucleus. Mass number A equals the number of protons plus neutrons in a nucleus.
4. Nuclear reactors should have a reproduction constant k about equal to 1, so that on the average each fission will result in one neutron that produces a subsequent fission. If k is less than 1, the reaction will die out. If k is greater than 1, there will be an exponential increase in fissions, and the reactor will "run away."
5. The cross section for neutron capture by $^{235}_{92}U$ is largest when the neutrons have small energies.
6. $U = 250$ MeV
7. $T = 5.70 \times 10^9$ K; $Q = 23.0$ MeV

Chapter 41

COMMON PITFALLS

1. False. Mesons and baryons are hadrons. Leptons are in a class by themselves.
2. Hadrons interact with one another via the strong interaction. Leptons interact with one another via the weak interaction. In addition, hadrons are composed of quarks, but leptons are truly elementary particles.
3. False. They have the same mass but opposite charge.
4. Once formed, a positron quickly meets an electron from the abundant supply of electrons in the matter in our universe to annihilate with.
5. False. Strangeness can change by ± 1 in weak interactions.
6. True
7. Six: up, down, strange, charmed, top, bottom. There are also six antiquarks.
8. True
9. A force between two particles arises from a back-and-forth exchange of a field quantum between the particles.
10. True
11. It unifies the electromagnetic and weak interactions into a single electroweak interaction.
12. False. It might be true if the word were "guts." But the word "GUTs" means that particle physicists are pursuing grand unification theories that unite the strong, electromagnetic, and weak interactions.
13. Theorists hope that grand unification theories will unify the strong, electromagnetic, and weak interactions into a single interaction.
14. False. The Big Bang refers to the enormous explosion that resulted in the birth of the entire universe.
15. Theoretical calculations show that the temperature of the blackbody radiation in the universe that has been expanding for about 10 billion years after the Big Bang will have cooled to the presently observed value of about 2.7 K.

TRY IT YOURSELF–TAKING IT FURTHER

1. The mass of Σ^0 is 1193 MeV/c^2, and the mass of Λ^0 is 1116 MeV/c^2, so by carefully studying the energetics and momenta of the reaction you might be able to determine if X was the more or less massive particle.
2. The paths of these particles curve in the direction associated with negative charges, given that the magnetic field was directed into the page.
3. The quantum numbers of this combination are different, and correspond to the Σ^0 particle.

QUIZ

1. True. To create an electron–positron pair, a photon has to have energy of at least $2m_ec^2 = 1.022$ MeV.
2. True
3. If only one photon were created, linear momentum could not be conserved.
4. Nucleosynthesis in stars cannot explain the observed abundance of helium, whereas the initially high temperatures of the Big Bang provide the reaction rates necessary to account for the present abundance.

5. Particle theory states that quarks must remain within the bounds of a hadron. As the distance between quarks increases, they experience larger and larger attractive forces. To separate quarks completely from each other would require an infinite amount of energy.
6. $d\bar{s}$
7. Reaction (b) is possible. Reaction (a) does not conserve charge as required.